Creative Packaging:
One-Piece Packaging Solutions

Paul Jackson

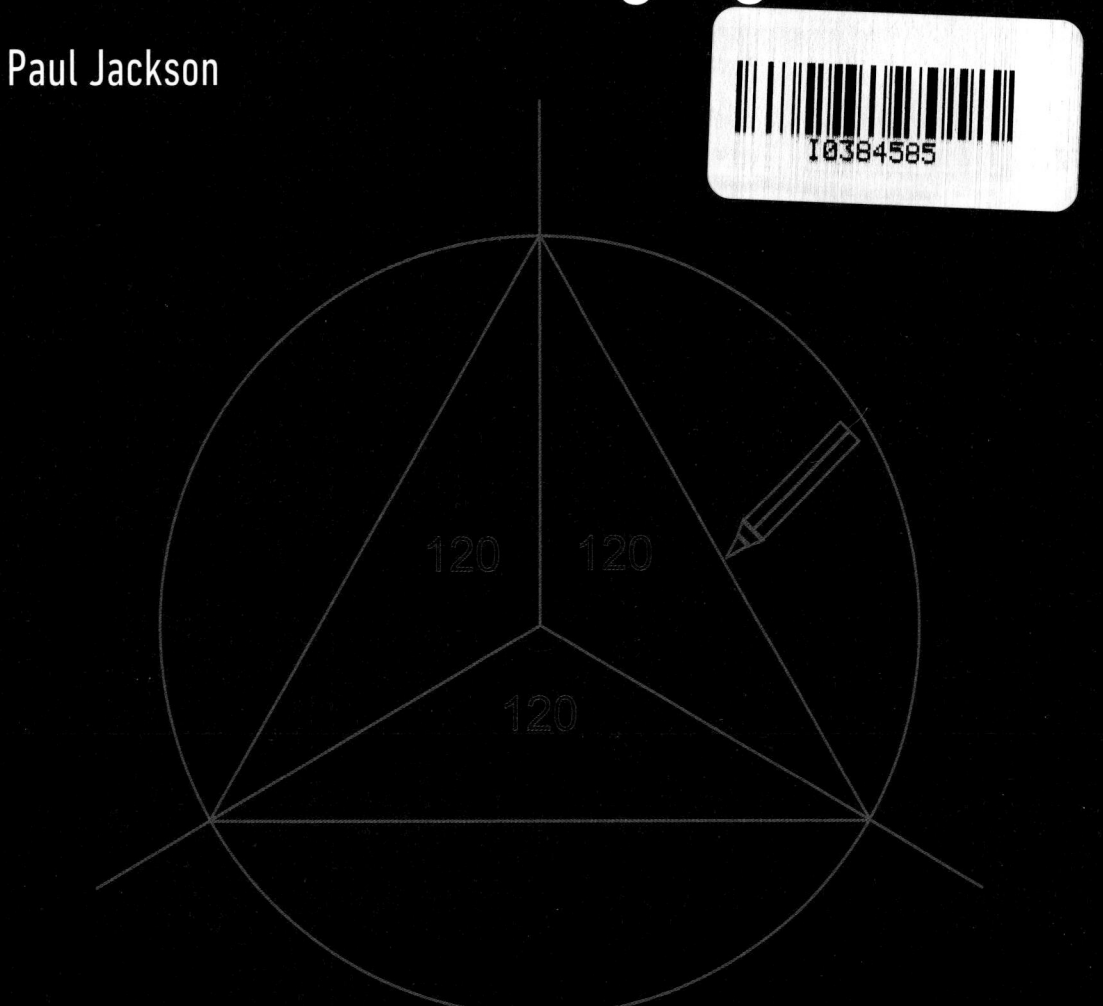

By One Of The Leading
Origami Professionals

sendpoints

INTRODUCTION

Like many manufacturing industries, the packaging industry is going through a period of profound change. Packaging with non-sustainable materials such as plastics and metals is in decline, replaced by sustainable materials, primarily cellulose fiber from trees, which make a wide variety of paper and boards.

Further, excessive packaging is now considered a needlessly wasteful use of materials and resources. It is expected that packaging designers will use the minimum amount of sustainable material, to preserve, protect and promote.

These new criteria make new demands on designers, who must now think minimally, think creatively and above all...think smart.

One way to think smart is to design packaging that is not the familiar cube or cuboid, but to reimagine the form. There are many books and online resources that offer "off-the-peg" packaging solutions to copy, but while these forms are familiar and efficient, they are also uninspired, even clichéd.

So, how does a packaging designer create original forms that are not just fanciful, but also strong and practical, and which use the minimum amount of material? This book has the answer!

The book explains my system of net design, which I first developed in 1984 (a "net" is the shape of the paper or board when a package is opened flat). I have taught it on perhaps 50 occasions to students of design in 7 countries, taught it to groups of professional designers and used it on countless occasions in my own professional practice as a paper artist-cum-engineer, not just to create packaging, but also to create point-of-sale displays, toys, models, decorations, exhibition designs and much more. My constant use and frequent teaching of this net design system has meant that I have come to understand through experience, what it can achieve, what is important to explain or ignore when teaching it, and how to teach it. This book is the summation of that knowledge.

The system is essentially a formula which, if followed precisely, will enable anyone to design the strongest self-locking net, using no glue and with the least material, for any given 3-D form. Thus, it becomes possible to design a one-piece net for (in theory) almost any 3-D form. It frees the designer from worrying HOW to create a net to thinking creatively WHAT to design. I have seen very many nervous students who have never made a 3-D form or a geometric drawing, create stunning work after only a few days.

I recommend that the book is read as though it was a novel, beginning at the start and reading it page by page to the end. In this way, the main points of the system will be learnt early and the subtleties later. It is also recommended that you make as much as possible by hand, not just work digitally with CAD software. This way, you will learn the physicality of your design and will better understand the finesses that can lift a good design to one that is faultless. For this reason, the book will take a little time to explain the technique of drawing nets by hand. You will be a better designer if you draw by hand, at least, in the initial phases of a project.

Used well and learnt fully, the system is powerful, creative and fun to use. It will enable you to design smart packaging – and much more – as your contribution to a future that is designed to be sustainable and beautiful.

Paul Jackson

CONTENTS

01 — PRELIMINARIES
P006

02 — THE BASIC METHOD
P022

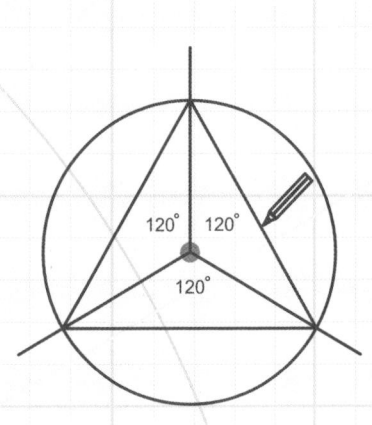

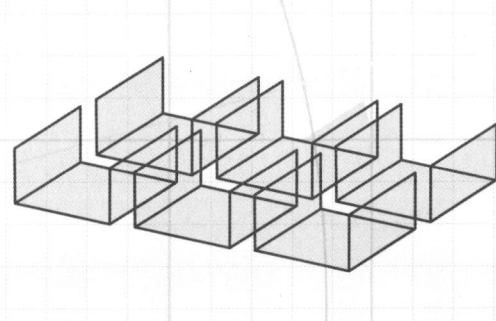

03 — CUBE DISTORTIONS
P098

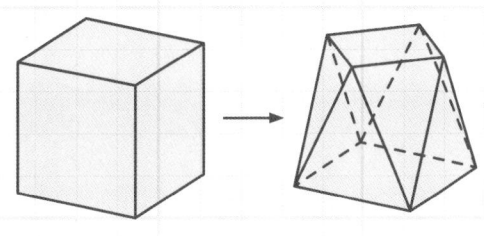

04 — LOCKS
P152

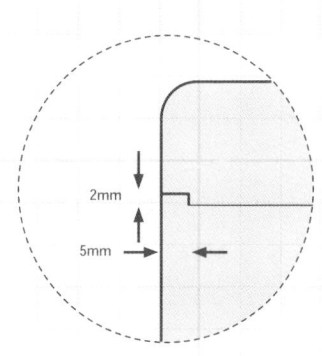

05 — CREATIVE PACKAGING... AND MORE
P168

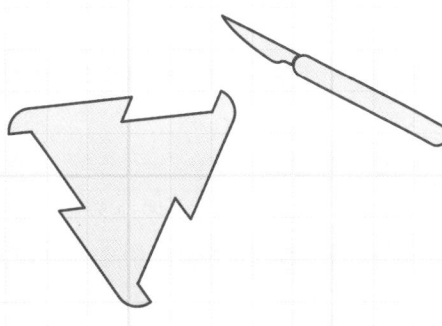

01

PRELIMINARIES

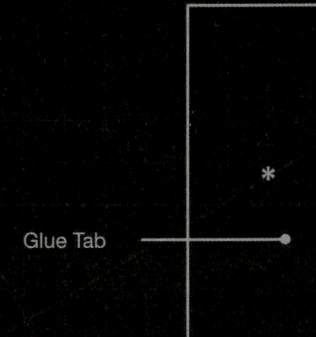

Lid Tab

Lid

Glue Tab

*

Tab

P008	**1.1 Which Card Should I Use?**
P008	**1.2 What Equipment Do I Need?**
P008	**1.3 Do I Need to Use Package Design Software?**
P010	**1.4 How to Draw on the Card?**
P014	**1.5 How to Cut and Fold the Card?**
P016	**1.6 Glossary**
P017	**1.7 How to Construct Polygons?**
P021	**1.8 Polygons**

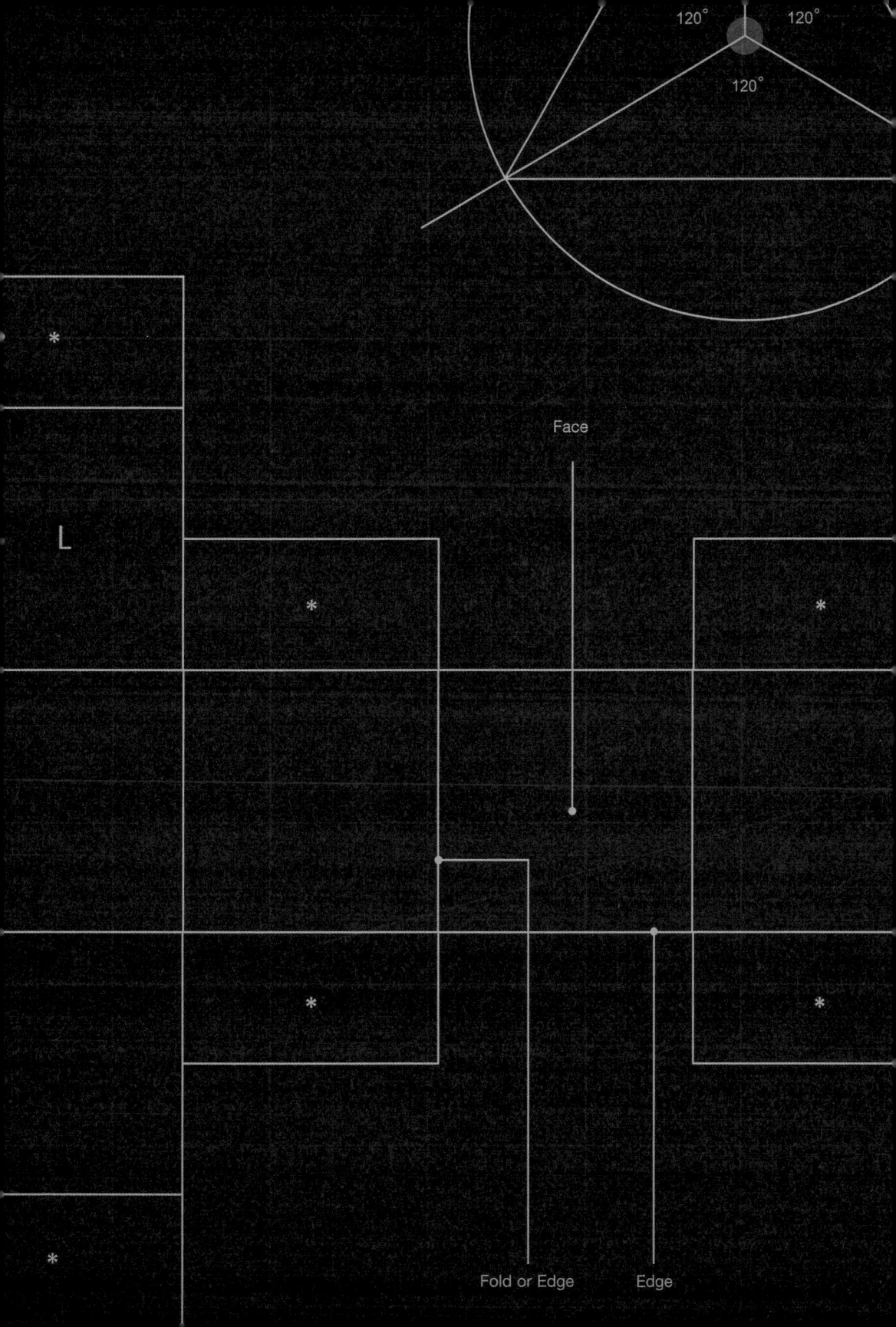

1.1 Which Card Should I Use?

Everything in the book has been made with 250gsm card. If you wish to copy what you see, use card in the range of 230–270gsm, but no thinner nor thicker. Further, do not use a card with a high gloss (shiny) surface. These cards are slippery and will not lock together well. It is recommended that you use matt card, as it will have more friction and thus more grip, when folded.

250gsm card is the ideal practice card because it is thick enough to have structural strength, but thin enough to cut and fold easily. However, in reality, few packages use material as thin as this. So, when you are satisfied that your net is correct, you should quickly switch to working with the material you wish to use for the final package.

Thicker materials often require that creases are moved a fraction of a millimeter to compensate for a turning circle, which occurs when the card is folded through an angle of 90-degrees, or whatever. Any movement of the crease lines should be done on a case-by-case basis, depending on the thickness of the material, the size of the package and the position of the crease lines in the net (side, lid, base…wherever). Although 250gsm matt card is the default card used in the book, any design described within its pages can be made with ANY thickness and type of card, even thick corrugated or double-wall corrugated card.

1.2 What Equipment Do I Need?

Before you begin, it is recommended that you assemble a collection of geometry equipment. Nothing on the list is expensive: you could probably buy two complete sets of equipment for the same price as this book. The one possibly expensive item is the self-healing cutting mat – a mat that can be cut on time and again, without becoming rutted or shredded. Large mats (A1, A0) can be expensive, but you can probably work well with one as small as A3. If these mats are beyond your budget, a sheet of inexpensive high-density card is an OK substitute, but you will need to keep replacing it. Under NO circumstances should you cut card on a table without using a surface to cut on.

1.3 Do I Need to Use Package Design Software?

In my opinion, the answer is no. At least, not at the beginning.

You are strongly recommended to make sketches and first prototypes by hand. In this way, you can quickly correct errors, get a feel for the shape, like how it sits on the table,

CARD & EQUIPMENT

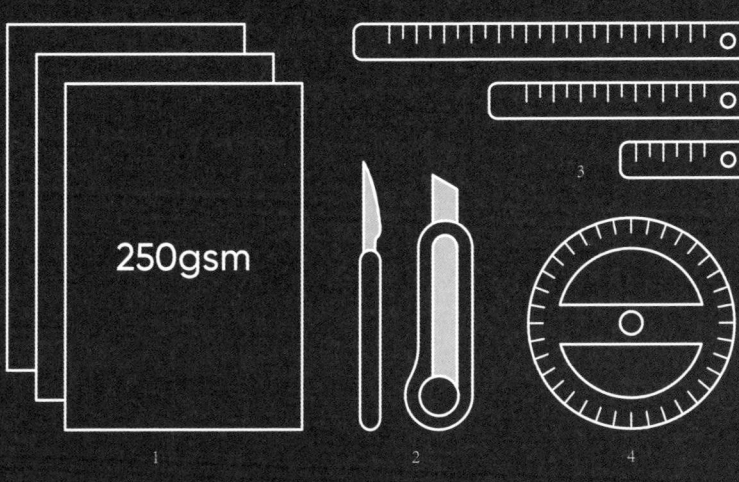

1. Stock of 250gsm matt card
2. Quality scalpel or craft knife, with many replacement blades
3. 15cm, 30cm and 50cm rulers
4. Protractor

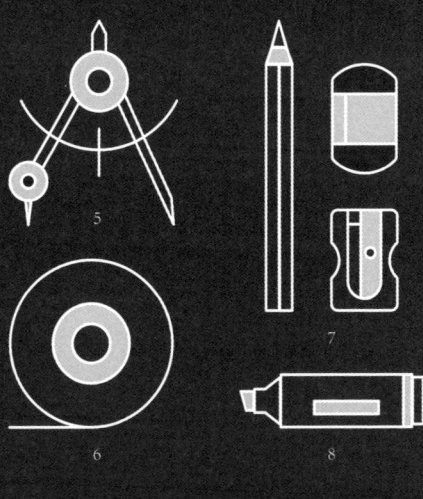
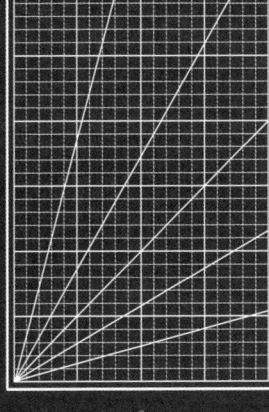

5. Pair of compasses
6. Masking tape (2cm wide)
7. 2H pencil, plus an eraser and a pencil sharpener
8. Marker pen
9. Self-healing cutting mat (or high-density card)

how it feels in the hand, and assess if the material is correct, if it can securely contain the product it will house... and more. Working by hand also means that you can work anywhere, anytime.

Of course, a time will come when you need to make a digital rendering of your net, perhaps also adding graphics. However, you should not move to this stage until you are confident that your hand-made net is accurate and you have solved all structural problems.

There are many, many CAD programmes available that can make drawings for engineers, product designers and architects, and they are all very capable of making accurate net drawings. Even graphic-based software such as Adobe Illustrator is very acceptable. The best advice is simply to use the programme you feel most comfortable with. If you are inexperienced at using CAD and simply feel it isn't worth a large investment of time to learn how to use the programme of your choice, it may be worth asking a friend to draw it for you. Packaging nets are usually very quick and very easy to draw for someone experienced with CAD.

The drawing can then either be printed out and cut and folded by hand, or be cut out using a plotter or laser cutting machine. Even at this late stage it is common to detect small errors in the drawing, so you should be prepared to make small corrections before declaring your design finished and 100% perfect.

1.4 How to Draw on the Card?

Some readers will be practiced at constructing accurate geometric drawings by hand, but in today's screen-based design environments, this skill is no longer taught to everyone.

If you are one of those people, please read this section with care. It will give you many essential tips and tricks for constructing perfect drawings.

There are two basic principles when drawing by hand:

1. Draw the Long Lines First

If you have ever taken an art class, when drawing a landscape, still life or figure, your teacher probably said, "Begin by putting in the major lines first, then the details later".

The same is true when drawing packaging nets: draw the long lines first, then draw smaller and smaller details as you progress through the drawing.

2. Use the Edges of the Card as a Reference when Measuring

Drawing lines that are parallel to each other, or perpendicular to each other, can be difficult. However, if instead of floating the drawing in the middle of the card, you use the edge of the card as a reference, making an accurate geometric drawing suddenly becomes easy.

Here is the method:

1.4.1

Here is the net for a simple cube which we want to draw.

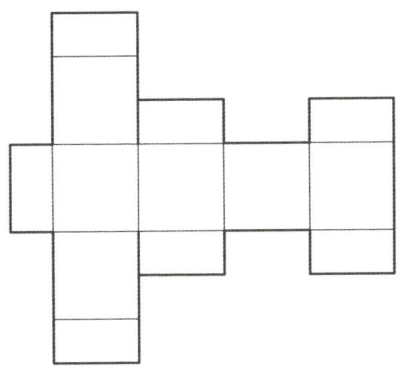

1.4.2

First, identify the longest lines, here shown in red.

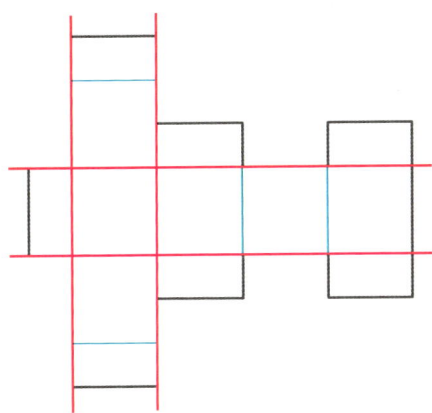

1.4.3

This is the configuration of the lines. The method shown here will describe how to draw these lines accurately.

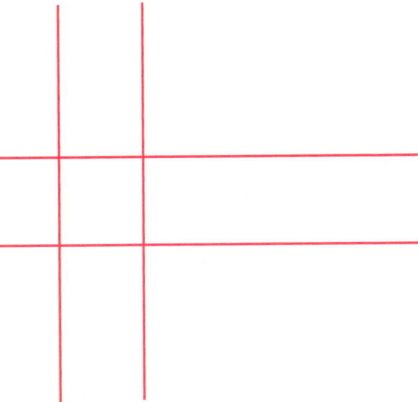

1.4.4

At the top left and top right corners of the card, measure the same short distance down the sides and make small lines.

1.4.5

Connect these short lines with a long line. The line will be parallel to the top edge of the card.

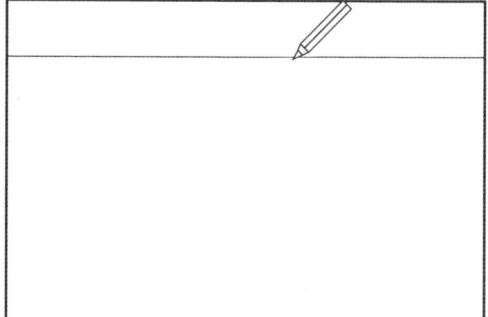

1.4.6

Similarly, make two marks down the left-hand edge of the card, some distance apart.

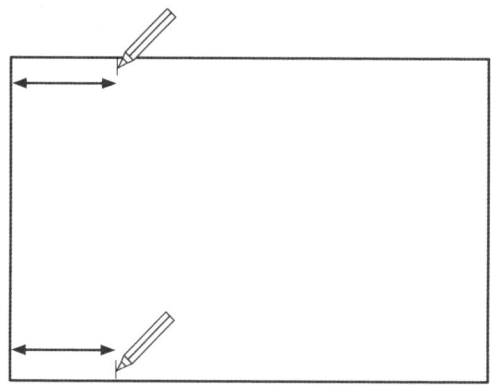

1.4.7

Connect these short lines with a long line.

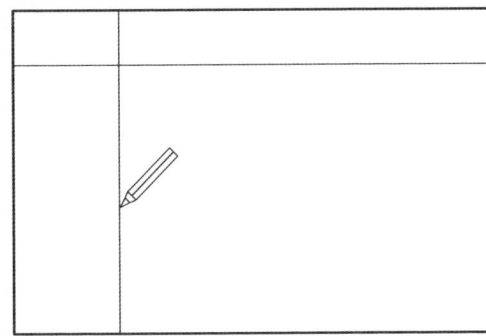

1.4.8

Make new measures as shown, then draw two new lines. These are your long lines on the net of the cube. You are now ready to make short lines.

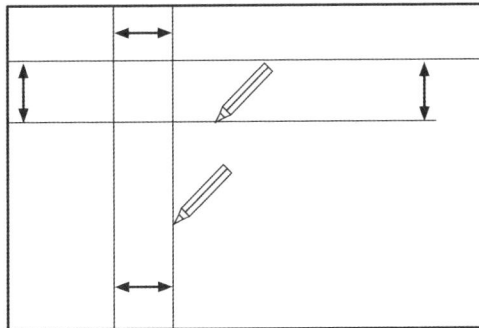

1.4.9

Make short lines as shown to complete the full net. To make them, use the long lines as a reference to locate the short lines. If you prefer, locate every line by measuring its distance from the edge of the card.

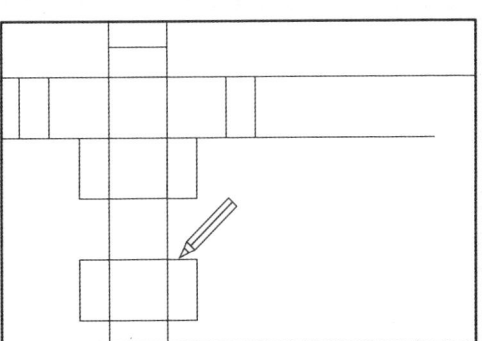

1.5 How to Cut and Fold the Card?

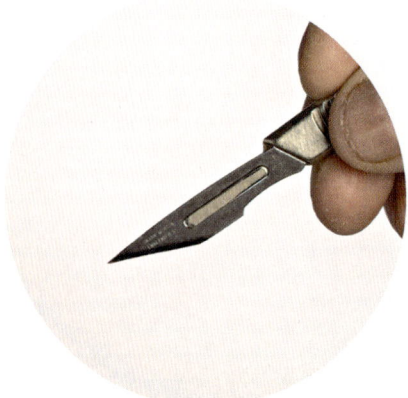

Knife

Cutting

When the net is drawn, the next step is to cut it out. This must be done with extreme precision, which is why you are recommended to draw with a sharp 2H pencil. This grade of pencil will create a pleasingly narrow line which will become the exact path of the cut. A softer or blunt pencil will create a wider line and consequently, an inaccurate cut.

If you are cutting the card by hand, it is important to use a quality craft knife or, better still, a scalpel. Avoid using the inexpensive "snap-off" craft knives, as they can be unstable and dangerous. The stronger, chunkier, "snap-off" knives are more stable and much safer. However, for the same price you can buy a scalpel with a slim metal handle and a packet of replaceable blades. Scalpels are generally more manoeuvrable through the card than craft knives and help you to better create an accurately cut line. Whichever knife you use, it is imperative to change the blade regularly.

A metal ruler or straight edge will ensure a strong, straight cut, though transparent plastic rulers are acceptable and have the added advantage that you can see the drawing beneath the ruler. Use a nifty 15cm ruler to cut short lines. Generally, when cutting, place the ruler on the drawing, so that if your blade slips away, it will cut harmlessly into the waste card around the outside of the drawing.

It is advisable to invest in a self-healing cutting mat. If you cut on a sheet of thick card or wood, the surface will quickly become scored and rutted, and it becomes impossible to make straight, neat cuts. Buy the biggest mat you can afford. Looked after well, it will last a decade or more.

When cutting, two separate cuts will sometimes meet, creating a corner on the net which is larger than 180-degrees (or written another way, the amount of card cut off at the corner is less than 180-degrees). This best done by starting a cut at the corner and cutting away from it, then returning to the corner and cutting the adjacent edge. In this way, the corner will be cut precisely. The alternative is to cut into the corner, which will invariably be less precise.

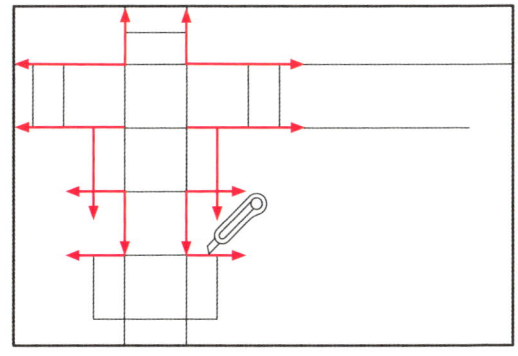

Here, the red lines show the cuts that need to be made away from a corner with less than 180-degrees of waste. Make each cut in the direction of the appropriate arrow. Not every edge to be cut is marked in red, because not every cut terminates in a dead-end corner.

Folding

While cutting paper is relatively straightforward, folding is less so. Whatever the method you use, the crucial element is never to cut through the card along the fold line, but by using pressure, to compress the fold line. This shall be done by using a tool. Whether the tool is purpose-made or improvised is a matter of personal choice and habit.

Bookbinders use a range of specialist creasing tools called "bone folders". They compress the card very well, though the fold line is usually 1-2mm or so away from the edge of the ruler, so if your tolerances are small, a bone folder may be considered inaccurate.

A good, improvised tool is a dry ball-point pen. The pen makes an excellent crease line, though like the bone folder, it may be a little distance away from the edge of the ruler. I have also seen people use a scissor point, an eating knife, a tool usually used for smoothing down wet clay, a fingernail (!) and a nail file.

But my own preference is a dull scalpel blade (or a dull craft knife blade). The trick is to *turn the blade upside down*. It compresses the card along a reliably consistent line, immediately adjacent to the edge of the ruler.

Although the edges in a regular cuboid box are all folded through an arc of 90-degrees, when making the folds, fold each edge through 180-degrees and press each double thickness strongly flat. Then, unfold the edge to 90-degrees. This is to ensure a crisp, straight, flat crease line edge between the planes of two adjacent faces.

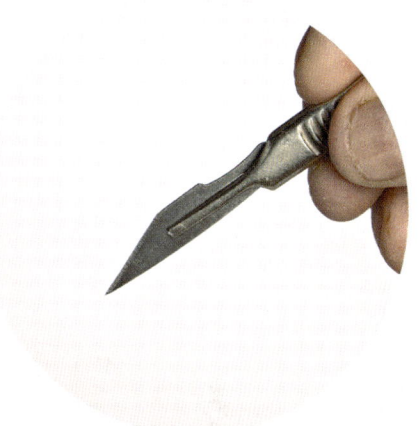

Blade upside down

1.6 Glossary

Like most specialist activities, structural packaging has a terminology all its own. When working through the book, refer to this section if you come across an unfamiliar term.

1.6.1 Box

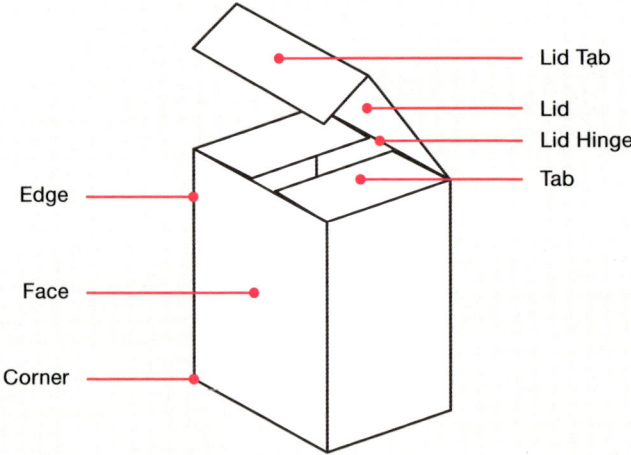

1.6.2 Net

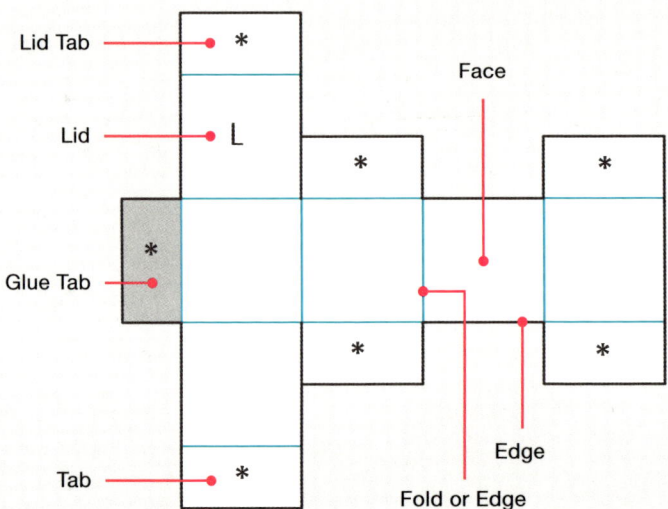

1.6.3 Valley and Mountain Folds

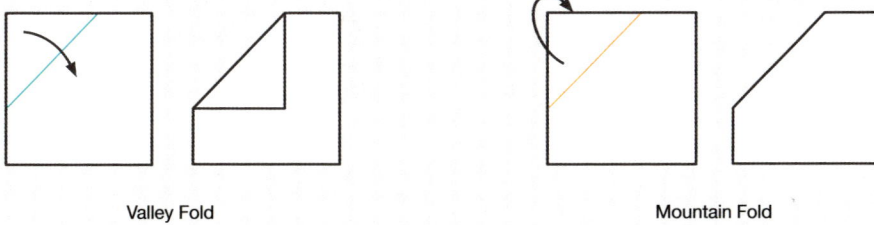

Valley Fold   Mountain Fold

1.7 How to Construct Polygons?

When constructing nets by hand, it is essential to know how to create accurate polygons. Even when constructing them using CAD software, it is still highly advantageous to know how they are made, beyond pressing the "Polygon" command key, then selecting the number of sides.

There are two basic ways to construct polygons. The first way is to construct one around a central point; the second way is to construct one from a line of a given length. They require different methods, so they are explained here, separately.

1.7.1 How to Construct a Polygon round a Point

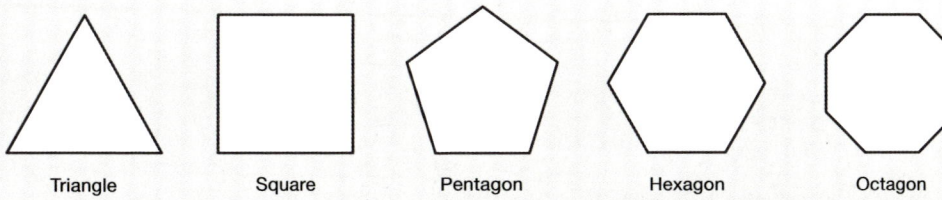

Triangle   Square   Pentagon   Hexagon   Octagon

1.7.1.1

Here is an equilateral triangle constructed around the central point. The method is shown as follows.

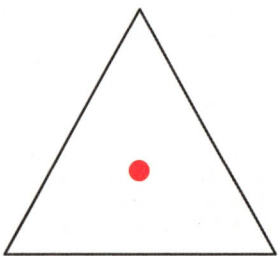

1.7.1.2

Make a dot on a sheet of card. Use a protractor and measure angles of 120-degrees around the dot. Why 120-degrees? This is because there are 360-degrees in a circle. An equilateral triangle has 3 interior angles equally spaced, so each corner below must be one third of 360-degrees apart, in other words 120-degrees.

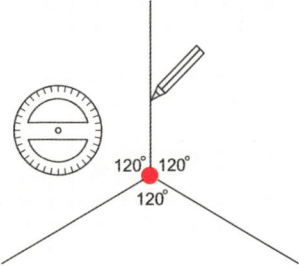

1.7.1.3

Center a pair of compasses on the dot and draw a circle of the appropriate diameter.

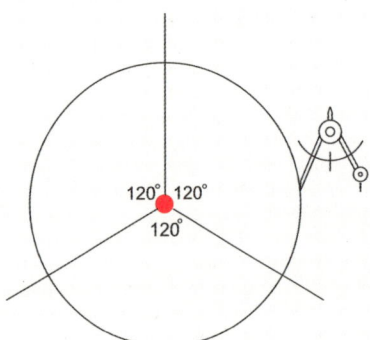

1.7.1.4

Draw lines connecting the places where the 120-degree lines and the circle intersect.

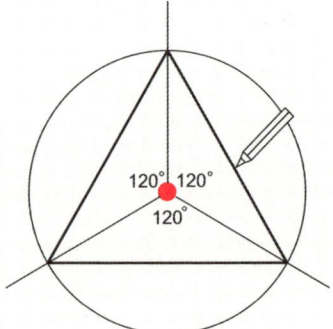

1.7.1.5

The equilateral triangle is completed.

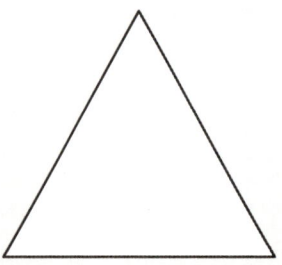

1.7.2 How to Construct a Polygon from a Given Edge

1.7.2.1
Here is the given edge.

1.7.2.2
We will create an equilateral triangle, allowing the given edge to be one of the 3 sides.

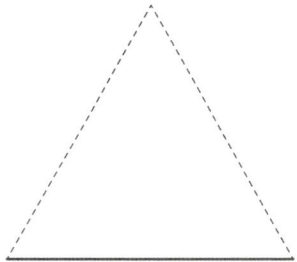

1.7.2.3
The technique is to first locate the central point of the polygon. Since each polygon has a different number of sides, the angle from the corners at the 2 ends of the given line to the central point, will be different for each.

To calculate this angle, we must first know the angle at the center. This can be found by dividing 360-degrees by the number of sides. Thus, it is 120-degrees for an equilateral triangle, 90-degrees for a square…and so on. Knowing the middle angle of the resultant isosceles triangle, means we can calculate the angles at the bottom corners of the triangle.

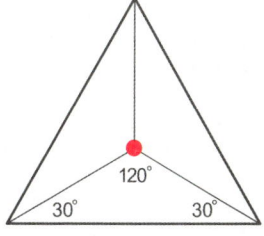
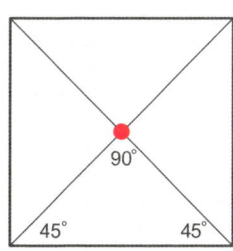
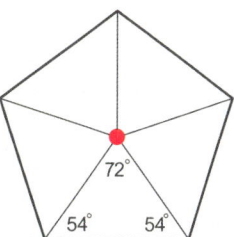
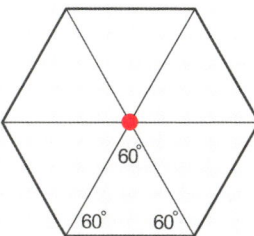
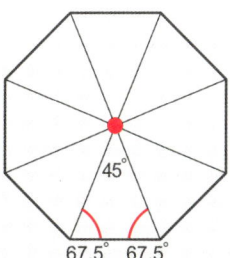

1.7.2.4

Imagine we wish to make a pentagon from a given edge.

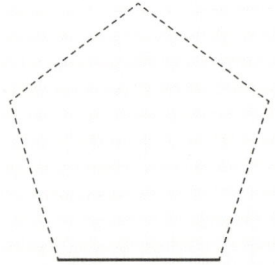

1.7.2.5

We can use the information above to calculate that the angle from a corner to the center point is 54-degrees. With a protractor, carefully measure these angles.

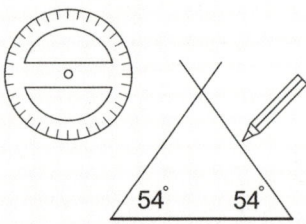

1.7.2.6

Where the two lines intersect will be the center point of the pentagon, where each angle will be 72-degrees. Use a protractor to measure 5 angles of 72-degrees.

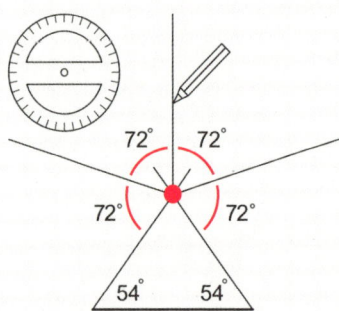

1.7.2.7

Use a pair of compasses. Place the needle at the center point and the pencil at one of the corners of the given edge. Draw a circle.

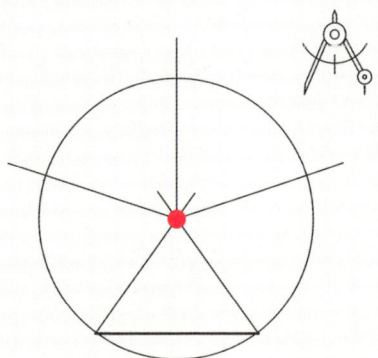

1.7.2.8

Where the circle intersects the 5 radiating lines, will be the corners of the pentagon. Connect them with a pencil.

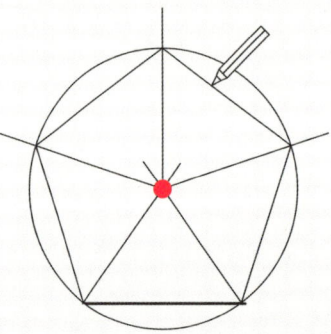

1.7.2.9

The pentagon is completed.

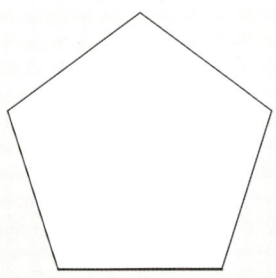

1.8 Polygons

Equilateral Triangle

(all angles and all edges are equal)

Isosceles Triangle

(two angles and two edges are equal)

Scalene Triangle

(all angles and all edges are different)

Right-angled Triangle

(one angle is a right angle)

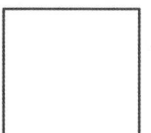

Square

(a four-sided polygon in which all angles and all edges are equal)

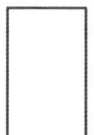

Rectangle

(a four-sided polygon in which all angles and opposite edges are equal)

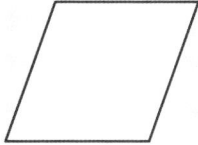

Rhombus

(a four-sided polygon in which opposite angles and all edges are equal)

Parallelogram

(a four-sided polygon in which opposite angles and opposite edges are equal)

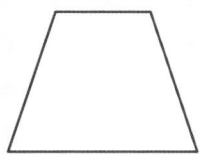

Trapezium

(a four-sided polygon with one pair of parallel edges and opposite angles total 180-degrees)

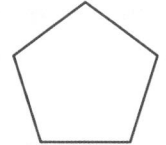

Regular Pentagon

(a five-sided polygon in which all angles and all edges are equal)

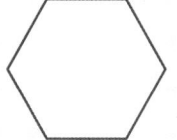

Regular Hexagon

(a six-sided polygon in which all angles and all edges are equal)

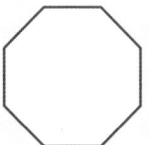

Regular Octagon

(an eight-sided polygon in which all edges and all angles are equal)

THE BASIC METHOD

P024	**2.1 The "Jackson Cube"**
P030	**2.2 From Six Modules to One Net**
P037	**2.3 Analyzing Why the Cubes Lock**
P041	**2.4 Generalizing the Net Theory**
P045	**2.5 Adding Tabs**
P050	**2.6 Including a Lid**
P056	**2.7 From Cubes to Cuboids**
P063	**2.8 Designing the Optimum Net**
P065	**2.9 Multi-tabbing**
P070	**2.10 Designing Nets with Corners that are not 90-degrees**
P079	**2.11 How to Lock Tabs with Angles of Less than 90-degrees**
P089	**2.12 Designing Nets for Faces with Corners Greater than 180-degrees**

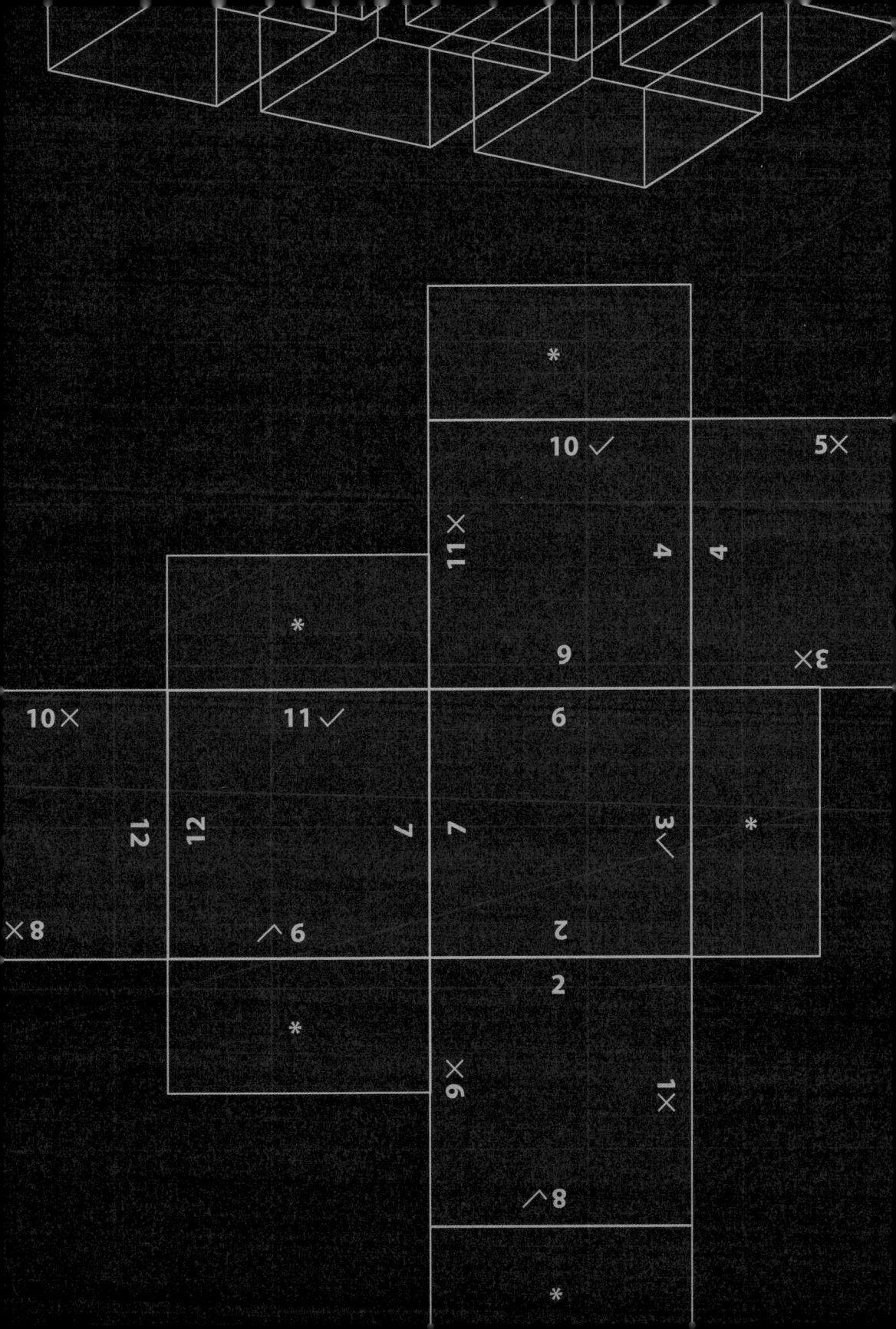

2.1 The "Jackson Cube"

This is a cube of my own design, dating from the early 1980's. It is so simple that it is considered the most basic of all modular origami (or "unit origami") designs, and whole families of designs can be derived from its locking pattern. If made well, it will hold together very strongly.

The Cube holds the key to understand the system of net design presented later in the book. A careful analysis of its structure will reveal how an one-piece net can self-lock very strongly, so rather than beginning with dry theory, it seems entirely relevant (and more fun!) to begin with this modular Cube.

2.1.1

In the corner of a sheet of 250gsm matt card, very carefully draw 9 lines, as shown.

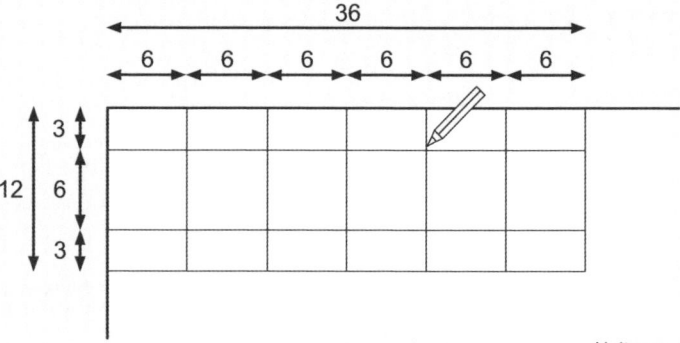

Unit: cm

2.1.2

This is the completed drawing. Note the 2 long blue lines. These lines must be creased with the back of your cutting knife.

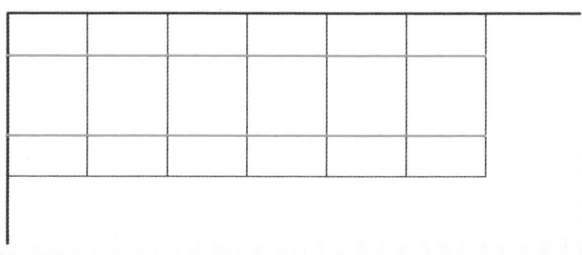

2.1.3

Make 6 vertical cuts, extending the cuts below the bottom horizontal line.

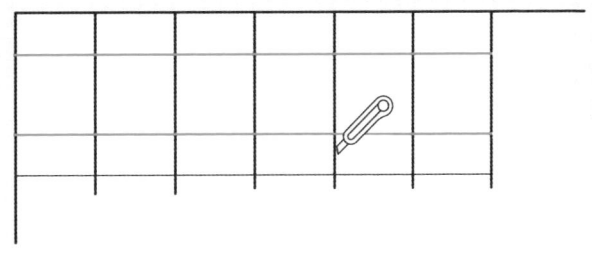

2.1.4

Make 1 long horizontal cut along the bottom line.

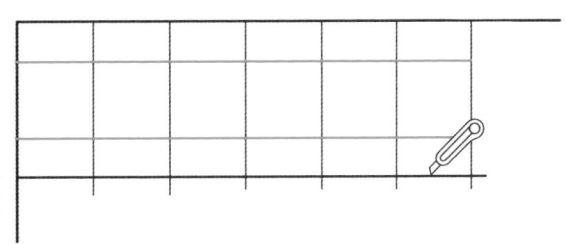

2.1.5

This final cut will release 6 rectangles of the card, each 12cm x 6cm.

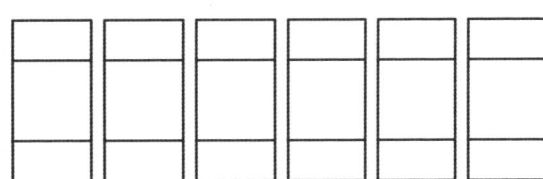

2.1.6

Crease each piece as shown.

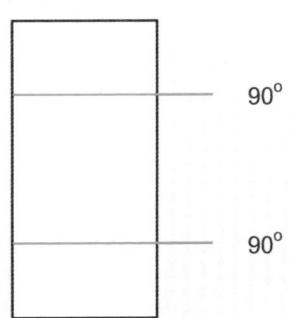

2.1.7

These are the 6 completed modules. Confirm that each crease is straight and sharp and the rectangles are standing upright.

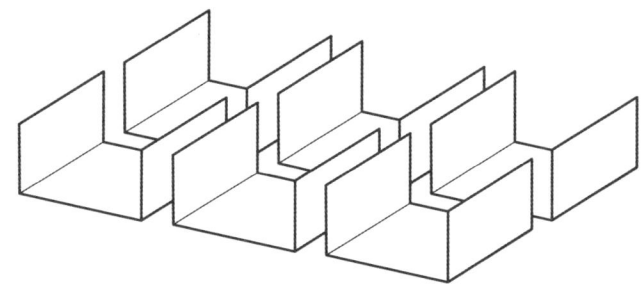

2.1.8

To interlock them, first lay 1 module on its back. Then, stand 2 modules upright and rest their rectangles on top of the first module.

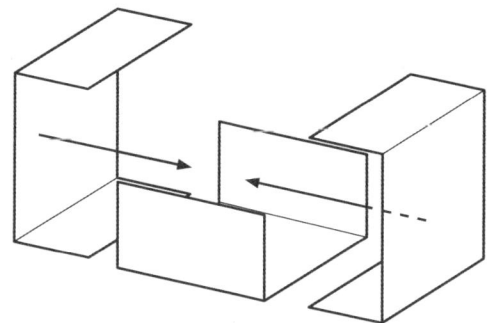

2.1.9

Now, slide in 2 more modules. Note how these new modules have their rectangles to the left and right, whereas the 2 modules in the previous step had their rectangles on top and bottom.

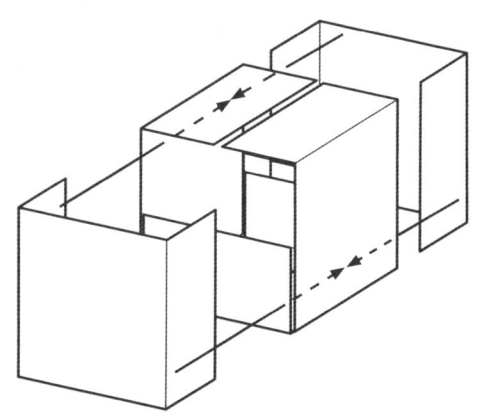

2.1.10

Finally, slide in the last module to close the top face.

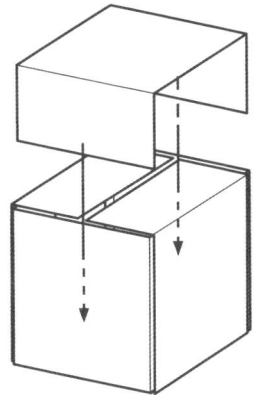

2.1.11

This is the completed Jackson Cube. If made well, it will be strong enough to play football with! If it does not hold together strongly, you have probably not made it accurately. Further, a matt card – rather than a shiny gloss card – will help the rectangles to grip inside the cube, holding the cube together with even greater strength.

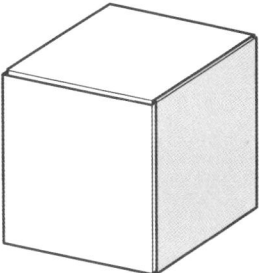

Conclusion

If you have not seen this Cube before, note how the 6 modules have 12 rectangles. The cube has 12 edges, so the rectangles that tuck inside are distributed one per edge. This regular structure and 3-D symmetry is why the cube is strong – any random arrangement of the pieces would hold together with less strength.

Many Jackson Cubes, different sizes and different colors, also multi-colored

2.2 From Six Modules to One Net

Having made the Cube from 6 modules, we can connect them together with tape, so that the cube can be opened flat and no module will fall to the floor. Clearly, this one-piece net will fold up to create a self-locking cube, because that was its original form!

Here is the method.

2.2.1

Cut 5 lengths of masking tape, each about the length of the cube's edge. This doesn't need to be done very accurately.

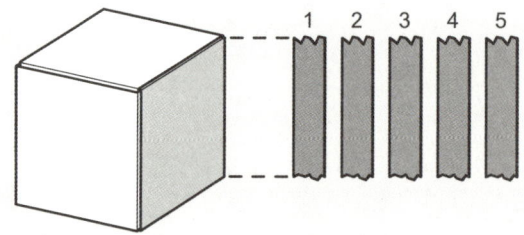

2.2.2

Now the puzzle!

Find a way to connect the 5 pieces of tape to the cube, so that the cube can be opened flat and nothing will fall to the floor. You will not solve the puzzle by looking at the cube, so just make a guess. Try anything! Take off pieces that are in the wrong place and move them to another edge. Keep moving and moving the pieces of tape. This may take a few minutes. Don't be afraid to make mistakes, just try things, learn, correct your errors and continue until you have succeeded.

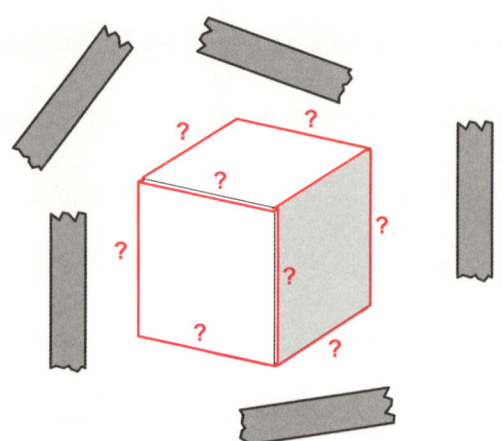

2.2.3

Open your cube and lay it flat.

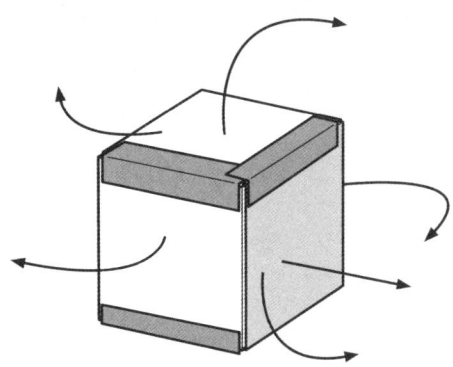

2.2.4

Let's analyze what you have made.

Your net will be a copy of one of these 22 drawings. These drawings represent the full set of possible nets when a cube is made from 6 square faces. There are no others.

Note that there are 11 different arrangements of squares, but each arrangement has an "a" and a "b", thus there are 22 drawings.

The "a" and "b" pairs show the rectangles at different places. Each rectangle in an "a" drawing appears where there is no rectangle in a "b" drawing...and vice versa. No rectangle in an "a" drawing appears at the same place in a "b" drawing.

The correct name for these rectangles is TABS. Note that there are always 7 tabs.

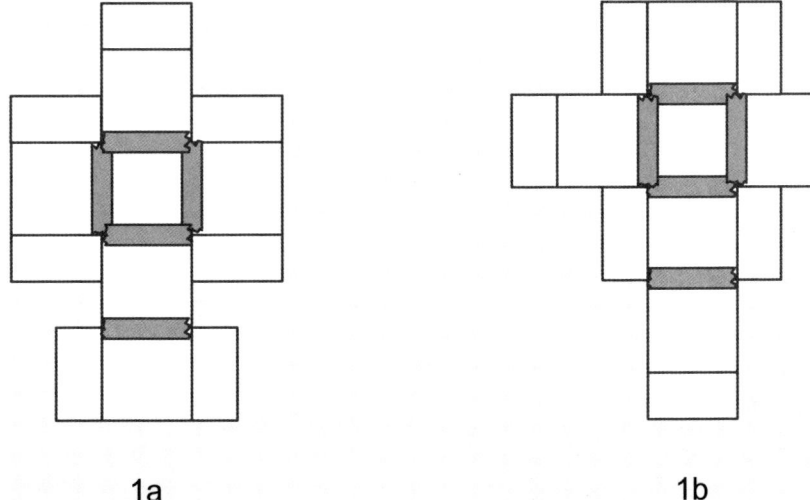

1a 1b

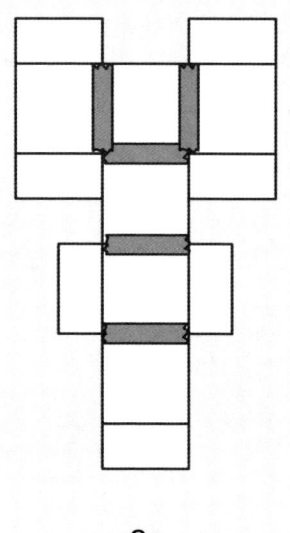

2a

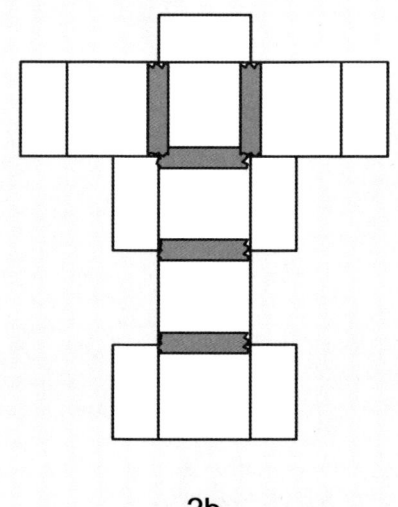

2b

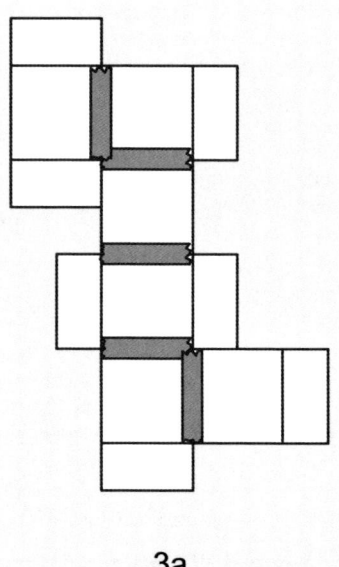

3a

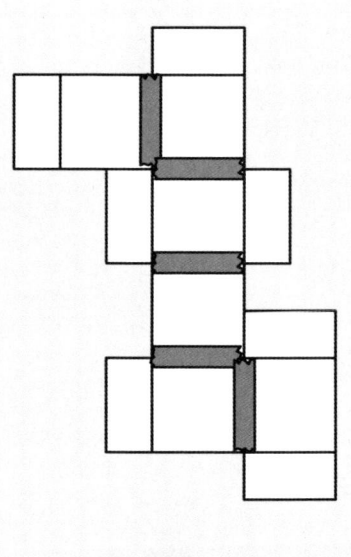

3b

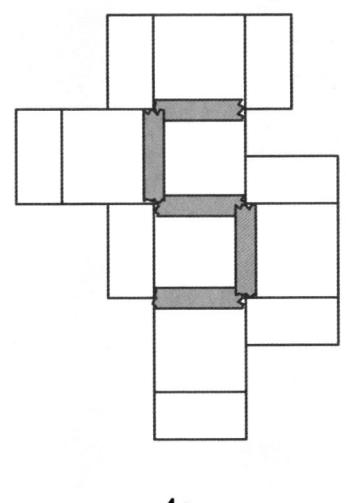

4a

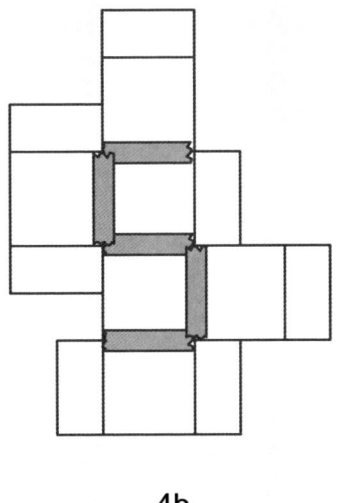

4b

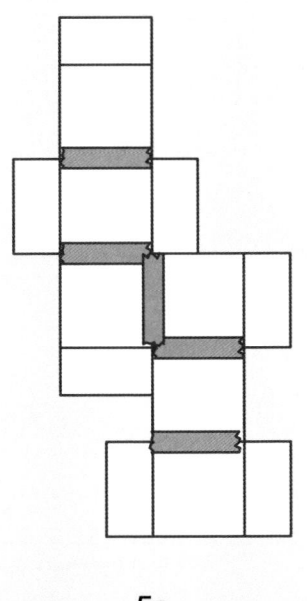

5a

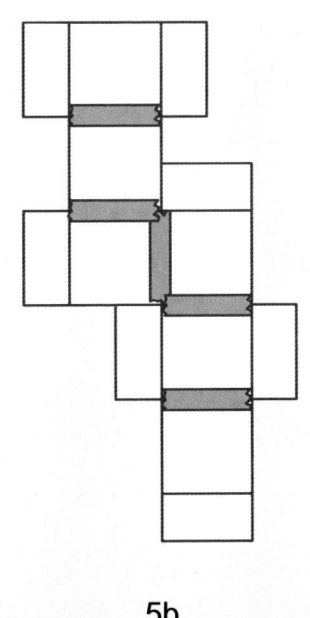

5b

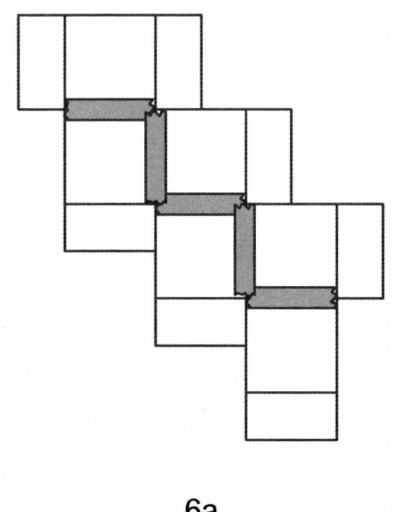

6a

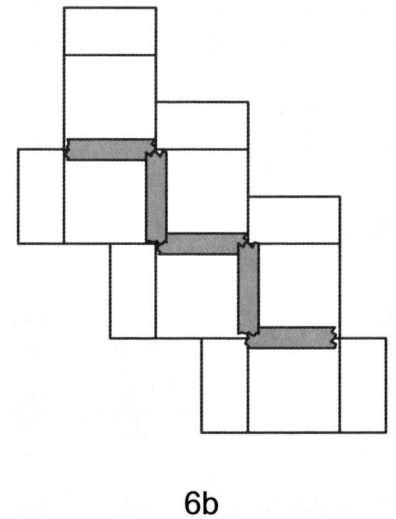

6b

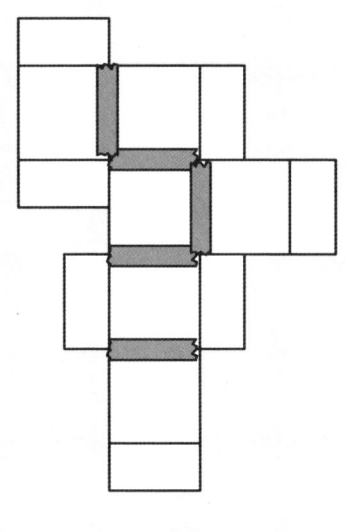

7a

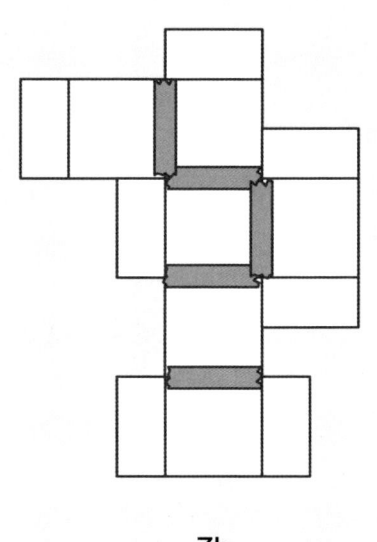

7b

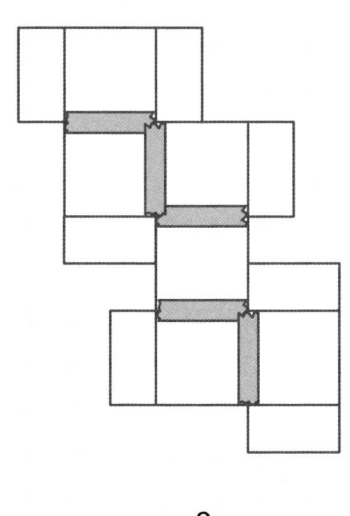

8a

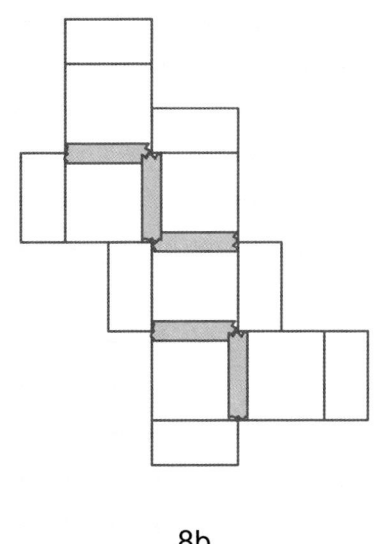

8b

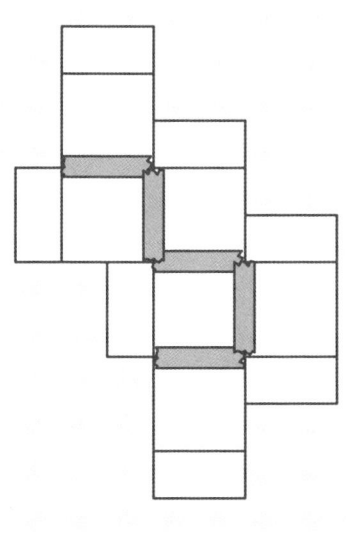

9a

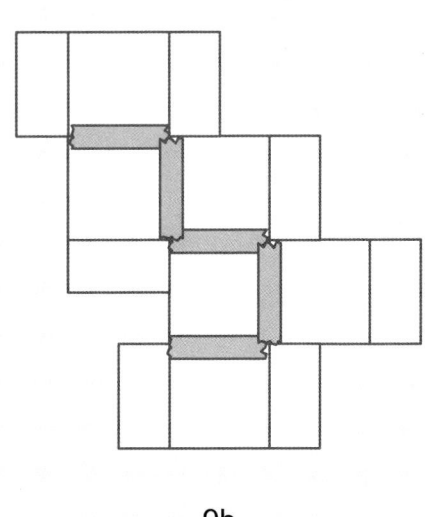

9b

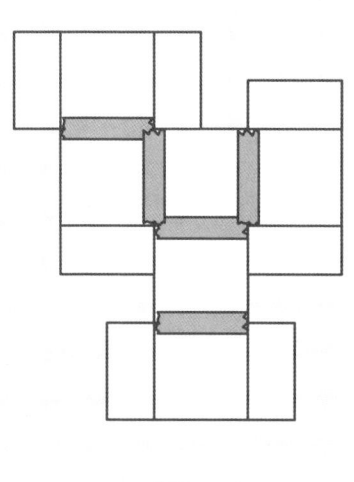

10a

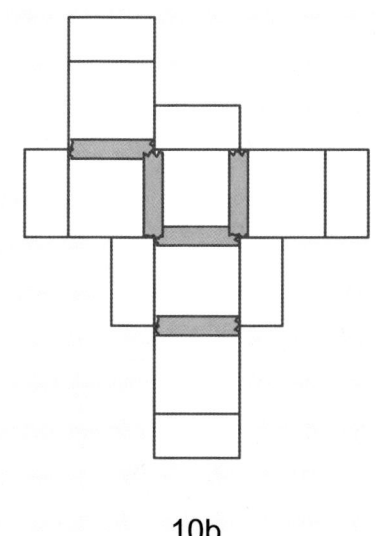

10b

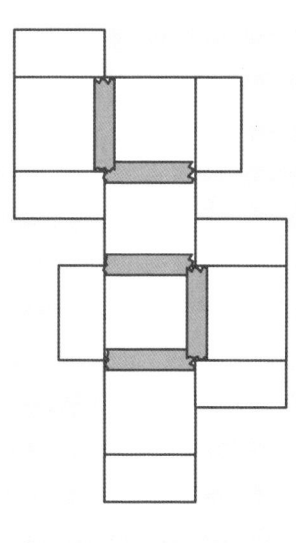

11a

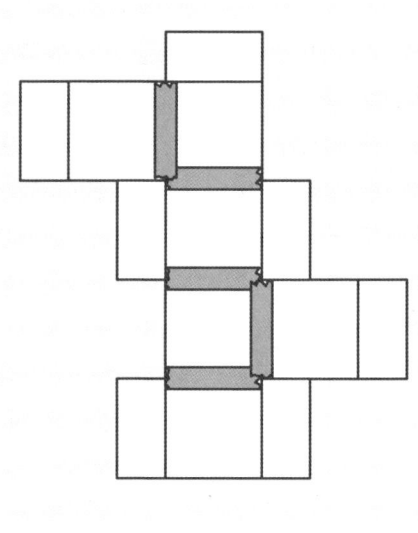

11b

Q: Can you see how the tabs are distributed in all the 22 drawings?

A: No 2 tabs are adjacent. They are always separated by ONE edge with no rectangle. Thus, the rectangles are on ALTERNATE EDGES.

This simple pattern is common to all the 22 possible nets. It must be the key to creating one-piece nets that lock strongly.

2.3 Analyzing Why the Cubes Lock

The 22 nets above are all different, yet they all can be folded up and locked to make identical-looking strong cubes. Understanding the reasons why these nets are successful will help us design other nets.

2.3.1 The Placement of the Tabs

2.3.1.1

Here are nets 4a and 4b. Note that the 6 square faces are in identical positions, but the tabs are in different positions.

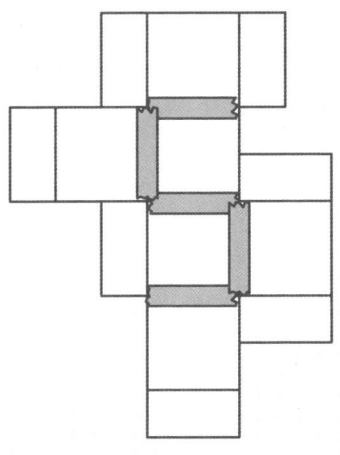
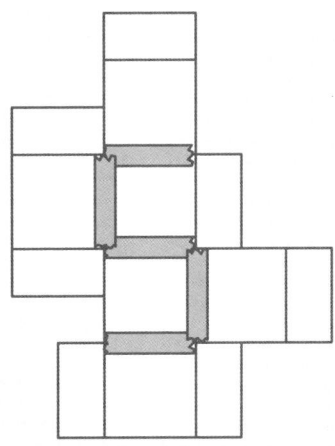

4a

4b

2.3.1.2

For clarity, the tabs are shown in strong colors, red and blue.

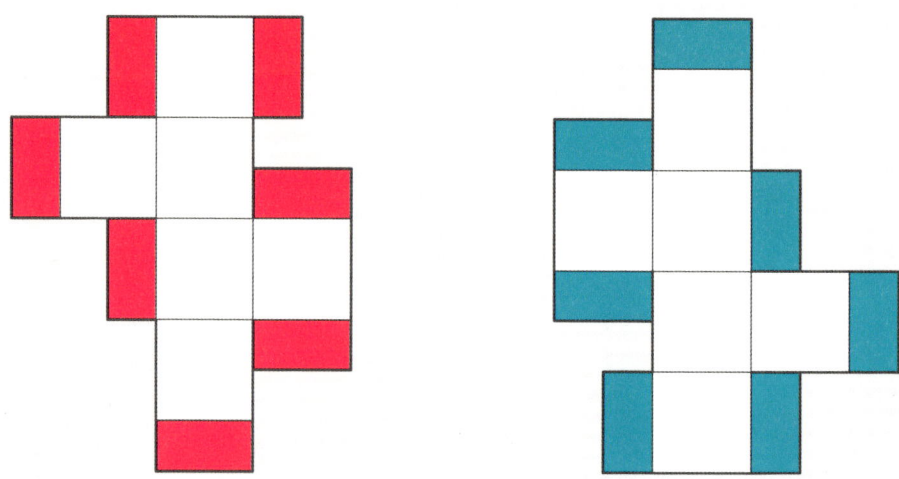

2.3.1.3

The colored lines show the edges where the tabs attach to the faces.

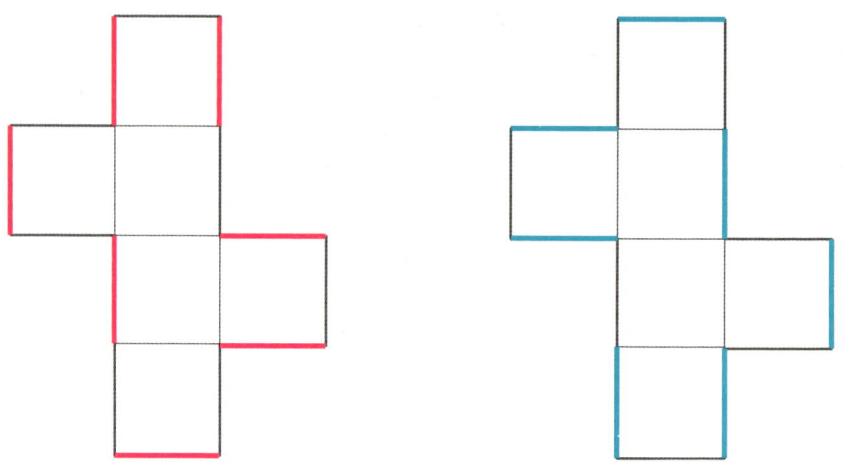

2.3.1.4

Note how the red and blue lines can both exist on one net! So we can see:

a. No edge is occupied by both a red and a blue tab.

b. All the edges of the net are colored.

2.3.1.5

If we replace the colored lines with ticks and crosses, we can see that the perimeter of the untabbed net has a regular "tick...cross...tick...cross...tick...cross" pattern. There are 7 ticks and 7 crosses. Every second edge has a tab and every second edge has no tab.

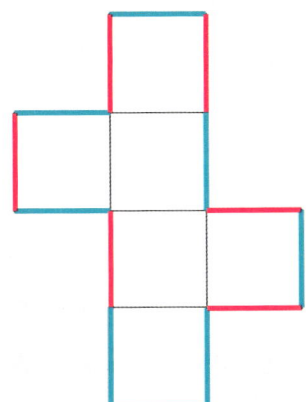

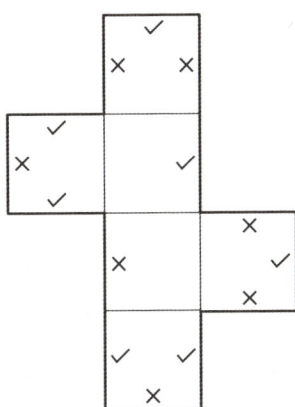

2.3.1.6

Here are the 2 nets. Note the "tab...no tab...tab...no tab..." pattern around the perimeter of both nets.

Now look at the 22 nets above in 2.2.4 and you will see the same pattern in them all.

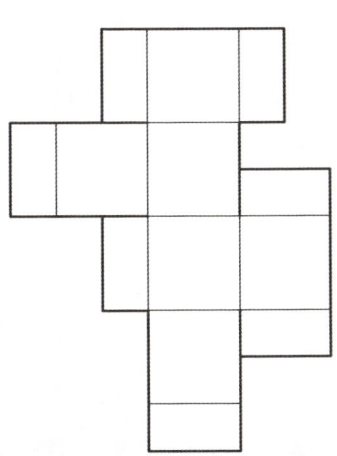

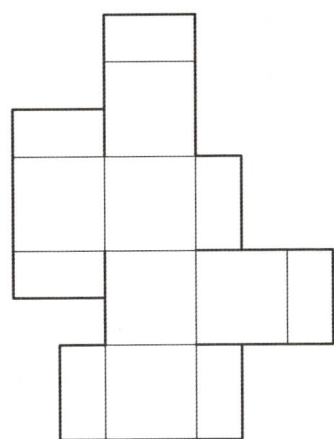

Conclusion

If the untabbed faces will be folded up to create an enclosed form, the form will lock together if the tabs are placed on every second edge.

2.3.2 The Shape of the Tabs

We now understand where the tabs must be positioned, but it is also crucial to understand the shape of the tabs.

2.3.2.1

Here we see a correct tab closing into the cube. Note how the shape of the tab exactly matches the shape of the face it is sliding behind. The face has 90-degree corners at the top, so the tab must also have 90-degree corners. This will ensure that the tab fits tightly into the cube.

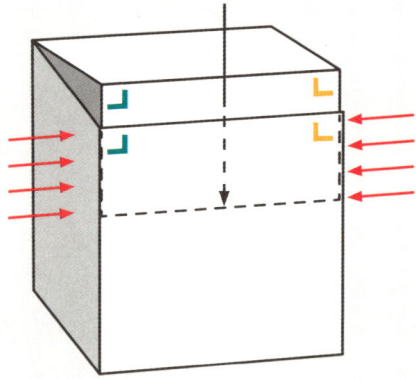

2.3.2.2

The length of the tabs is also important. Notice how the length of each tab is half of the length of the face. If you made the example earlier in the Chapter, the faces were 6cm x 6cm and the tabs were 3cm x 6cm, or half a face. Thus, the grip of each tab is maximized, creating a stronger lock.

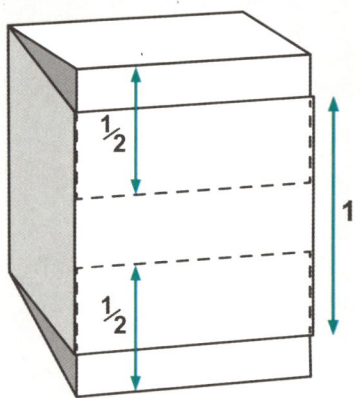
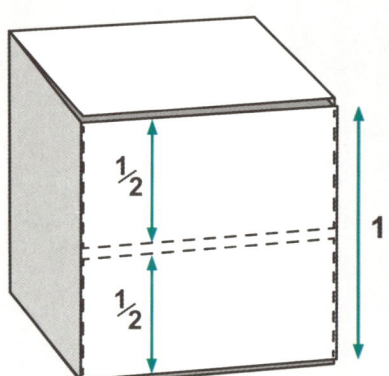

2.3.2.3

The first drawing shows a perfect tab. It is half the length of the edge and has 90-degree corners.
The second drawing has a correct 90-degree tab, but it will not lock strongly because it is too narrow.
The third drawing shows a large tab, but the angles are not 90-degrees. It will not grip inside the cube and lock it together. If all the angles were made this way, we could use glue to lock the cube closed. However, if we make the net and the tabs smartly, we do not need to use glue! This is a good design.

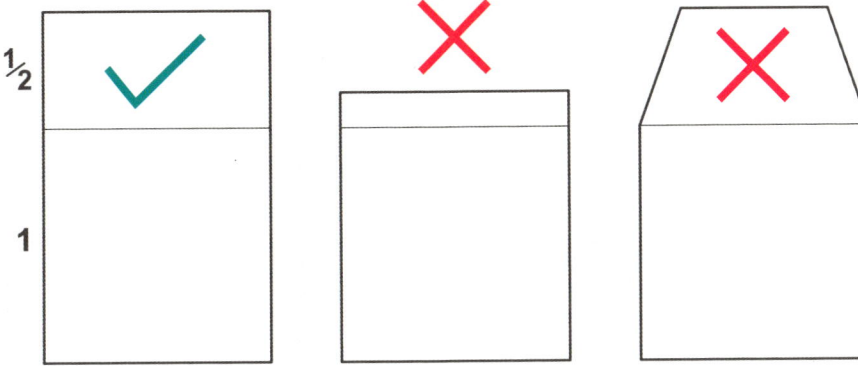

2.4 Generalizing the Net Theory

We have seen how the Jackson Cube generates 22 possible one-piece nets in 11 pairs.

However, the process we went through to create a net was specific to that cube and would not work if we wanted to create a net for, say, a pyramid.

So, we need to generalize the theory, expanding it from a specific example to a method that will work for any 3-D solid.

Here is how we can generalize the method.

2.4.1

On a sheet of card, measure 6 squares, as shown. They do not need to be made with accuracy.

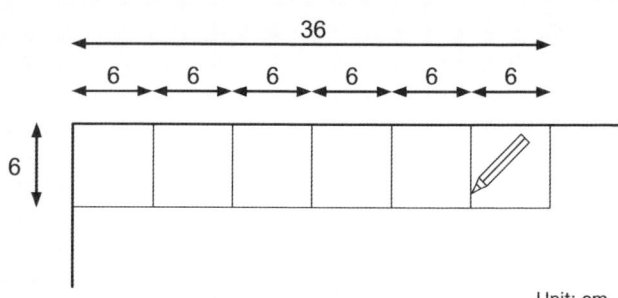

Unit: cm

2.4.2

Make 6 cuts, as shown.

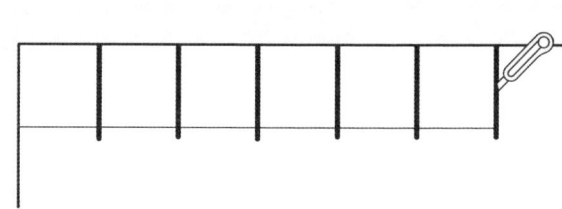

2.4.3

Make one long cut, to release the 6 squares.

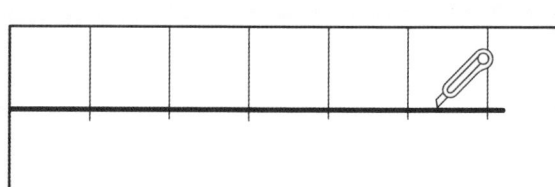

2.4.4

Here are the 6 squares.

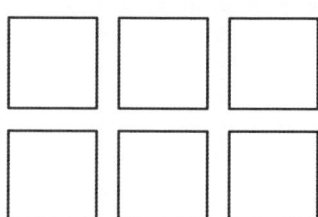

2.4.5

Cut 12 pieces of tape, one is for each edge of a cube. Each length of tape is the approximate length of the side of the square.

2.4.6

Apply one piece of tape to each edge of the cube. This does not need to be done carefully.

2.4.7

Use a marker pen or a very soft pencil to write numbers VERY CLEARLY on the cube.

The numbers are written in pairs 1, 1...2, 2...3, 3...4, 4 across each of the 12 edges, in any order, from 1 to 12.

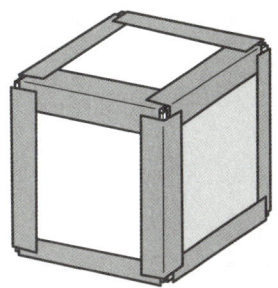

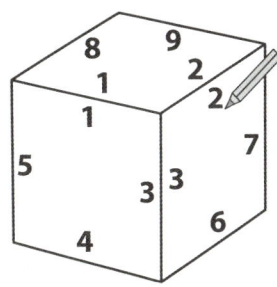

2.4.8

This is one example of how the numbers could be written. There are actually hundreds of millions of ways to do this!

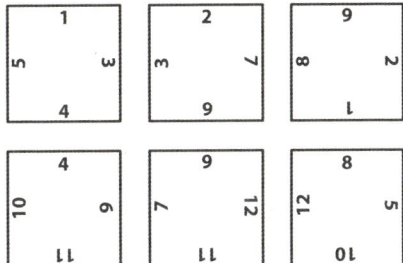

2.4.9

Think about it:

However the number pairs are placed together, the resulting configuration would fold up to make a cube. In this example, the numbers 4, 6, 2, 7 and 12 have been placed together. If we were to fold this net, the other number pairs would also come together when the cube was 3-D; so, 1 would touch 1, 3 would touch 3...and so on.

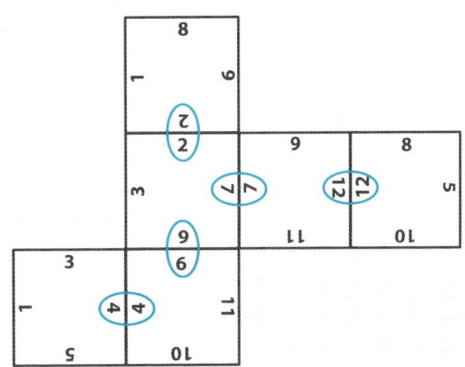

043

Mini Task

Experiment with your numbered squares and draw the possible unique nets they can make, without creating mirror nets or rotated nets. How many can you find?

2.4.10

There are 11 possible configurations of the squares that will fold up to create a cube, excluding mirror nets or rotated nets.
The numbers themselves are irrelevant. What is relevant is that when the 6 squares are laid flat, or when the cube is folded to 3-D, those like numbers touch.

2.5 Adding Tabs

We have learnt how to create configurations of squares that will fold up to create a cube. However, these squares will not hold together because they are missing the 7 rectangles that tuck inside and lock the faces together.

This section will show you how to add those rectangles.

2.5.1

Here is the net we made in the previous section, but it could be any one of the 11 possible unique nets.

Starting at any edge, write a tick very clearly next to the number. On the next edge, write a cross next to the number, then a tick, then a cross, then a tick, then a cross…and so on…around the complete perimeter. There will be 7 ticks and 7 crosses.

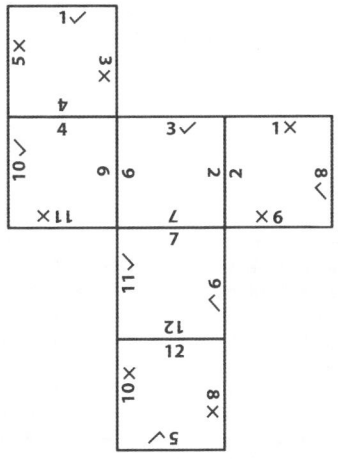

2.5.2

On a sheet of card, measure and cut out 7 rectangles, each 6cm x 3cm. The 6cm length is the length of the edge of the cube. 3cm is half of that length.

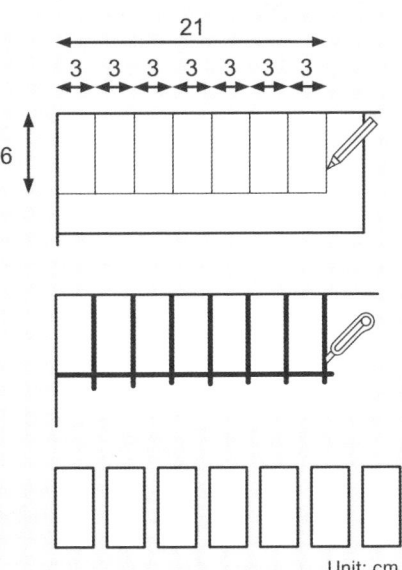

Unit: cm

2.5.3

Join the 7 rectangles to the squares ALONG THE TICKED EDGES ONLY! Notice how each rectangle joins to a unique number.

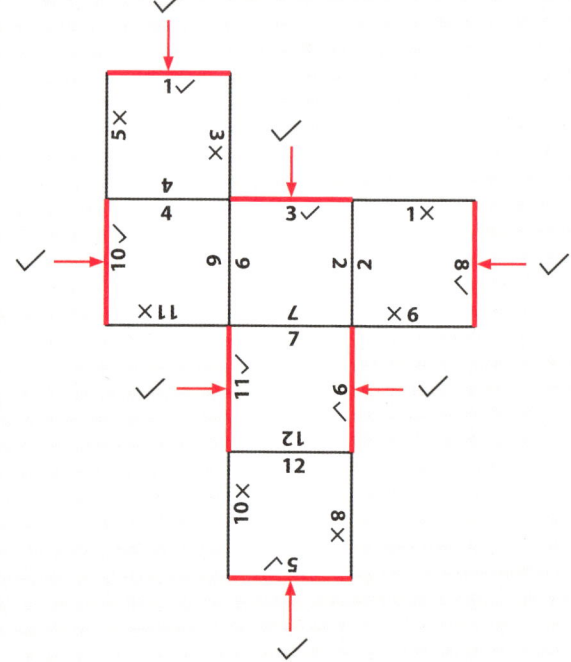

2.5.4

This is the final collage of squares and rectangles.

The rectangles have a special name. They are called TABS. It is the tabs that lock a 3-D solid together, but they are usually inside, unseen and unappreciated.

Of course, you could also place all the tabs on the crossed edges. However, you can never mix placing them on both ticked and crossed edges.

If you have made a mistake or wish to create another net, move your squares around and re-write the ticks and crosses in the appropriate places on the new perimeter.

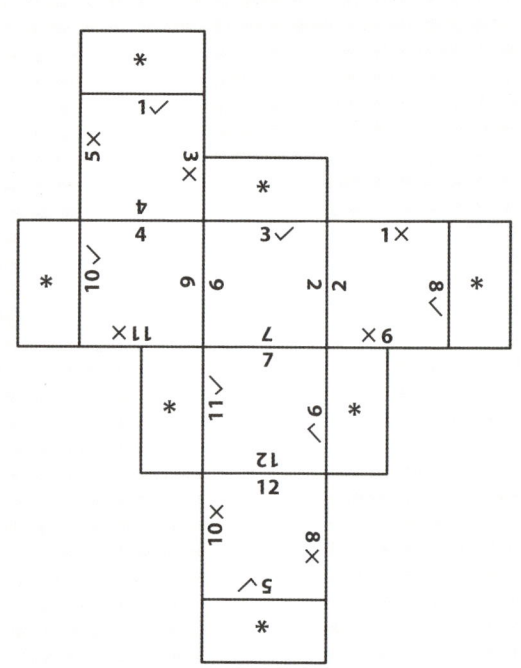

2.5.5

When you are happy that everything is correct, the complete cube, with tabs, can be made accurately from one sheet of card.

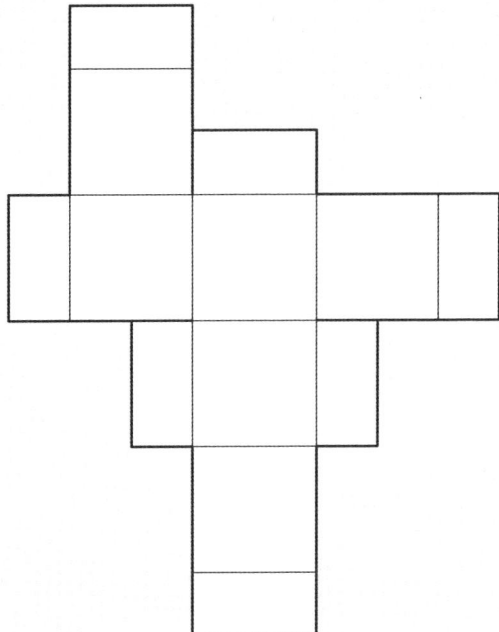

Rough-looking collage of faces and tabs, numbered and taped

Perfectly-made one-piece cube

2.6 Including a Lid

Any package needs a means to be accessed. It needs a lid that can be opened and closed.

However, if you are designing self-locking solids for purposes other than packaging – for example, as a point-of-sale display – then you may not need a lid.

But assuming you need a lid, how can you design one into your net? This section will give you the answer.

2.6.1

Here is a cube with a face cut loose on 3 sides. The fourth side is a fold that connects the face to the cube box. The loose square is a lid. Of course, it needs tabs both to lock it to the box and to prevent it from collapsing inside.

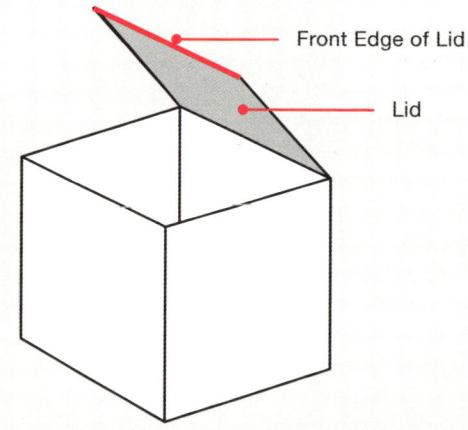

2.6.2

There will be 3 tabs associated with the lid. 1 tab connects to the lid along its front edge, while 2 more tabs connect to the box, under the left and right edges of the lid. When the lid is closed, the tab on the front edge holds the lid to the box and the supporting tabs prevent the lid from collapsing into the box. Thus, the 3 tabs all secure and support the lid square.

The package designer must create these 3 tabs in his specific configuration around the perimeter of the net.

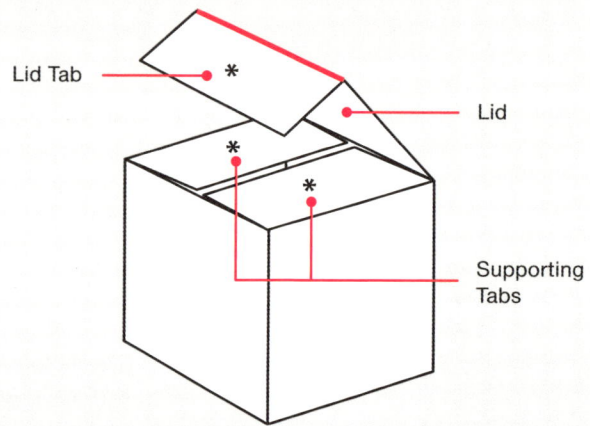

2.6.3

Here are the 11 unique nets that create a cube. Look again at the drawing in 2.6.1 and note that the lid has 3 raw edges on the perimeter of the net. Thus, only squares that have 3 edges on the perimeter of a net can be candidates for a lid. Those squares with only 2 edges on the perimeter cannot become lids.

The squares with 3 edges on the perimeter are marked in gray. Note that some nets have 2 such squares, whereas others have 3 or 4. The red edges denote the front edge of a lid.

2.6.4

Here is one of the 11 nets. 3 of its squares have 3 edges on the perimeter and could become lids.

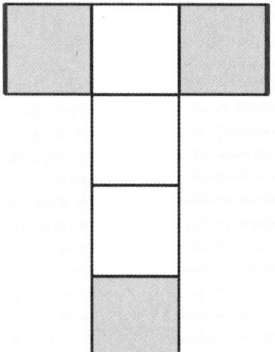

2.6.5

Let's choose this square to be the lid.

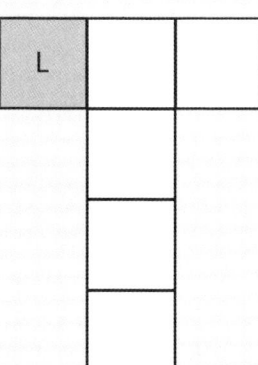

2.6.6

Write a tick on the front edge of the lid – the front edge is between the 2 other lid edges that are on the perimeter of the net.

This edge is the most important edge on the perimeter, because it must have a tab – see the drawings in step 2.6.2.

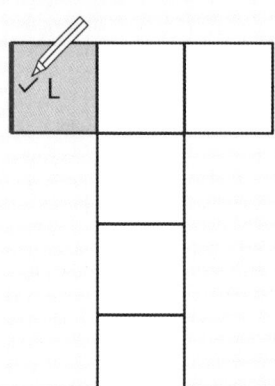

2.6.7

Join a tab to that front edge.

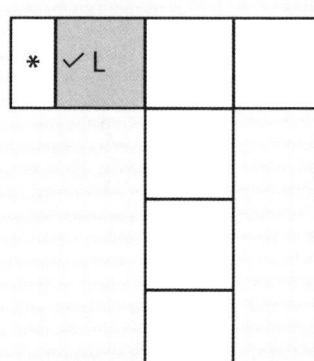

2.6.8

Mark the remaining edges of the net with alternating crosses and ticks, so that no 2 ticks are adjacent and no 2 crosses are adjacent.

2.6.9

Join tabs to the ticked edges. This is the complete and correct net for a cube which has a lid in the position shown.

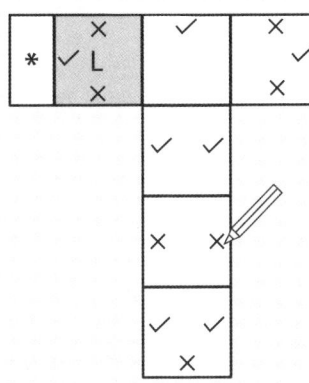

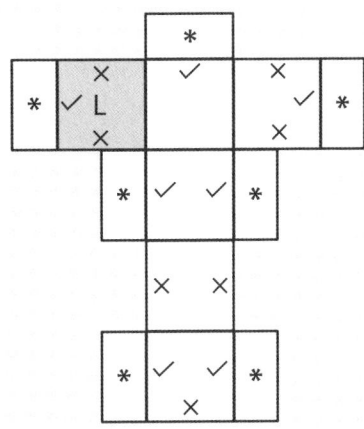

2.6.10

It is also possible to create a lid from the square at the bottom of the same net. As before, identify the front edge of the lid and mark it with a tick.

2.6.11

Continue the tick-and-cross pattern around the perimeter and add tabs to the ticked edges. This is the complete net.

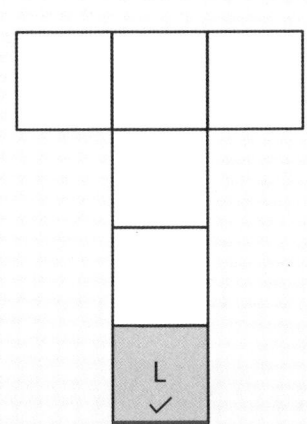

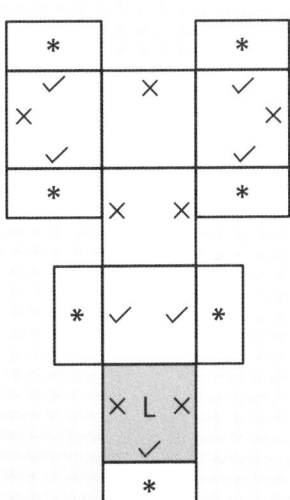

2.6.12

Given the choice of 11 unique nets and 31 possible lids, the choices even for a package as simple as a cube can be somewhat bewildering. This is where other factors become important, such as the strength of the net, its size and the way the printing wraps around the folds.

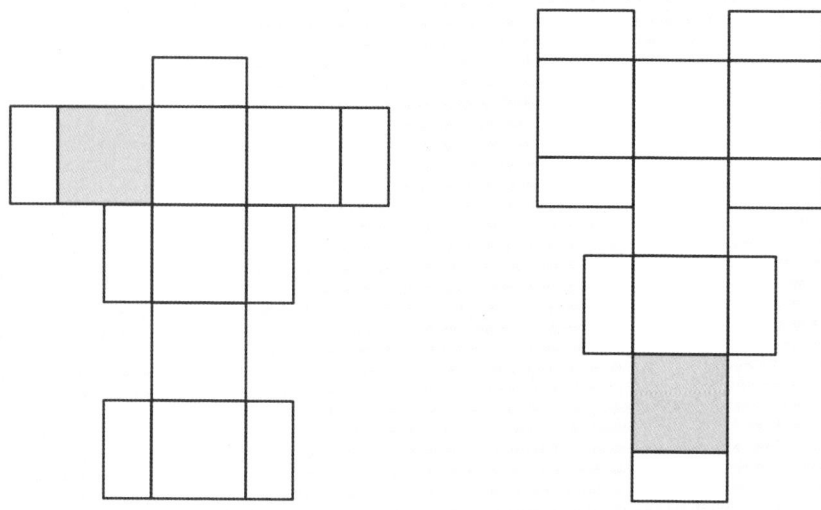

Conclusion

Cubes are perhaps the simplest of all package designs to create, so the next section will begin to develop the net design system explained above for other, more complex, 3-D forms.

Cube with open lid, showing tabs

2.7 From Cubes to Cuboids

A cube is made from 6 square faces. All its 12 edges are the same length.

A cuboid is similar to a cube, except that its faces are not usually square, but rectangular. It has 3 different rectangles, 2 of each. Thus, while there is only 1 cube, there are an infinite number of cuboids of different rectangular lengths and proportions.

This section shows how the net design system explained above can be adapted to work with cuboids.

2.7.1

The drawing shows one of the 11 nets for a cube. Each face is a square and all the squares are the same size. Note the gray square at the top of the net.

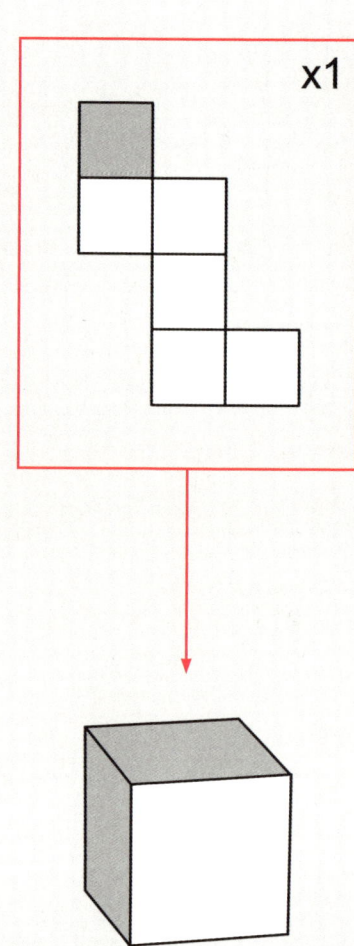

2.7.2

Now consider a cuboid. It has 3 different rectangles, each of which can be horizontal or vertical on the net. Thus, if a cuboid net is drawn that is the equivalent of a net for a cube, the gray rectangle at the top of the net can be any one of the 3 different rectangles, each in 2 possible positions. Designing the net from top to bottom to copy the configuration of the cube net, we are presented with 6 choices for different nets, whereas there was only 1 choice for a cube!

This means that whereas we had 11 unique nets for a cube, we now have 66 possible nets (11 x 6) for any cuboid. The complication isn't in the increased number of possible nets, but in the designing of the tabs.

Here is the method. It begins the same as for the cube. Basics are basics!

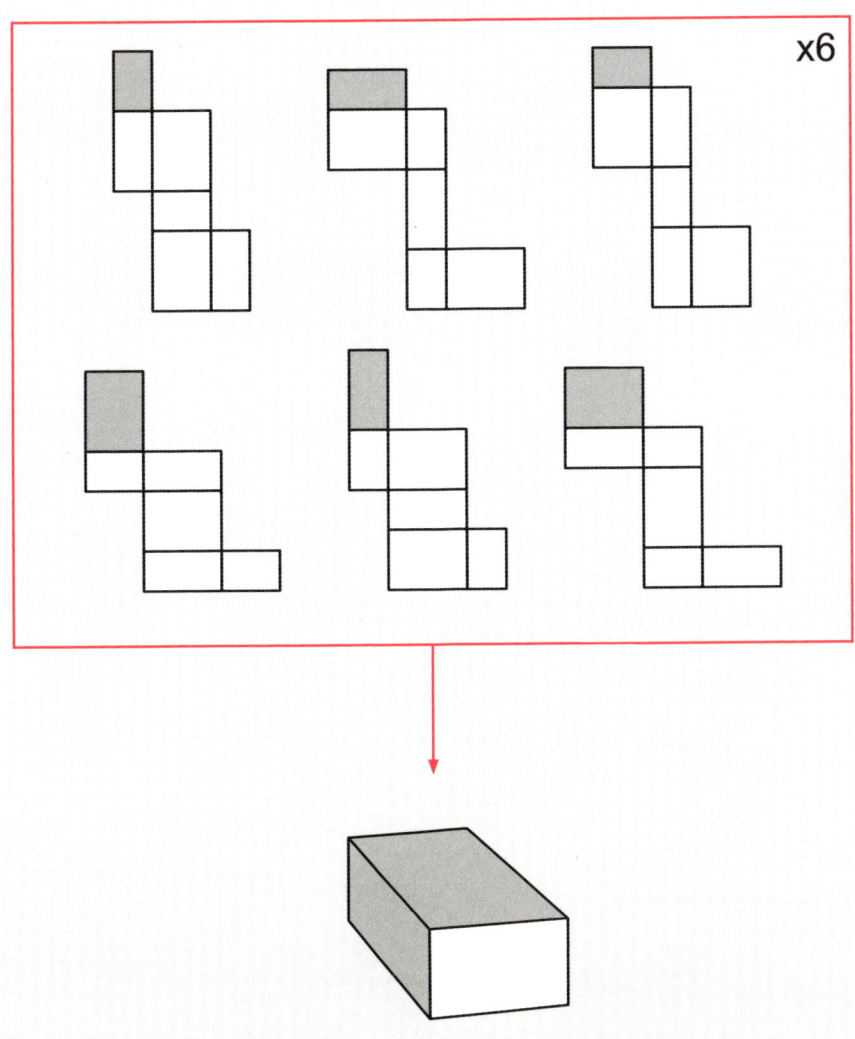

2.7.3

Here is an 8cm x 6cm x 4cm cuboid. Measure the rectangles and make 2 examples of each.

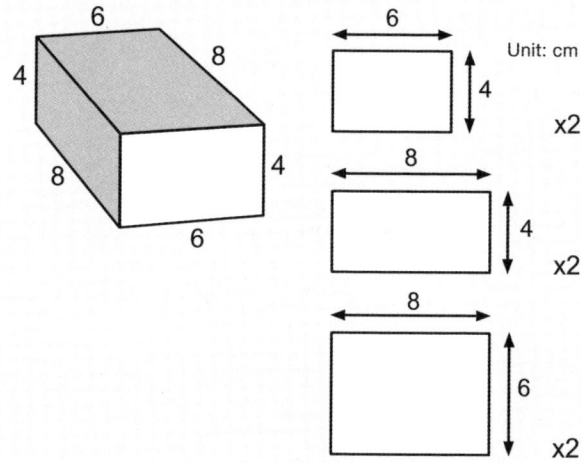

2.7.4

Tape the rectangles loosely together and add clearly written pairs of numbers randomly across the 12 edges.

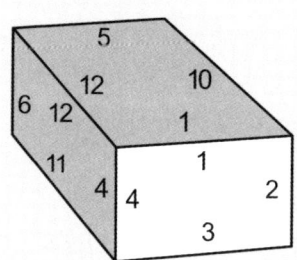

2.7.5

Separate the faces.

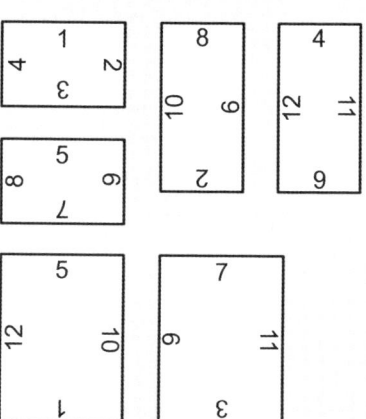

2.7.6

There are 66 possible ways to create a net, joining 5 pairs of like numbers together. Here is 1 net chosen at random.

We need to add a lid and then tab the perimeter.

The lid is the rectangle at the bottom right. The front edge of the lid is marked in red. Write a tick clearly next to the number, then add a tab. Continue to add a cross, then a tick, then a cross, then a tick…and so on…around the remainder of the perimeter. There will be 7 ticks and 7 crosses.

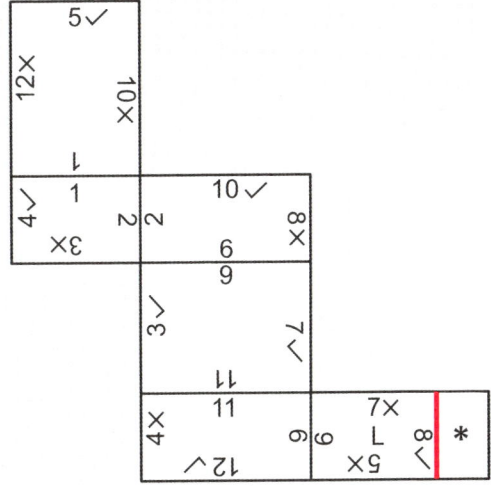

2.7.7

We are now ready to add the 6 remaining tabs. The cuboid has 3 edge lengths…8cm, 6cm and 4cm…so the DEPTH of the tabs will be HALF of these lengths… namely 4cm, 3cm and 2cm.

We must learn the shape of each tab, one at a time. There is no "one size fits all", like a cube. This puzzle can take just a few minutes, or many minutes, depending on the complexity of the net.

Let's look first at edge "5-tick", at the top of the net. The tab attached to it will fit under edge "5-cross" when the cuboid is folded into 3-D. The distance behind edge 5-cross, to 7-cross, is 4cm. Thus the depth of the tab will be HALF of 4cm, which is 2cm. Thus, the tab attached to 5-tick will be 2cm deep.

The same is true for the tab attached to 7-tick. It will also be 2cm deep, because the face beyond edge 7-cross is the same 4cm to edge 5-cross.

Cut out the tabs and tape them to the growing net.

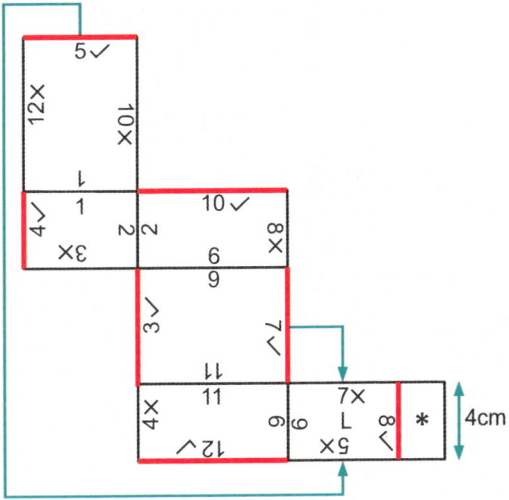

2.7.8

Here we can see the tabs added to edges 5-tick and 7-tick.

2.7.9

The tabs attached to 10-tick and 12-tick will fold under edges 10-cross and 12-cross. The distance between these 2 edges is 6cm, so each tab will be 3cm deep (half of 6cm).

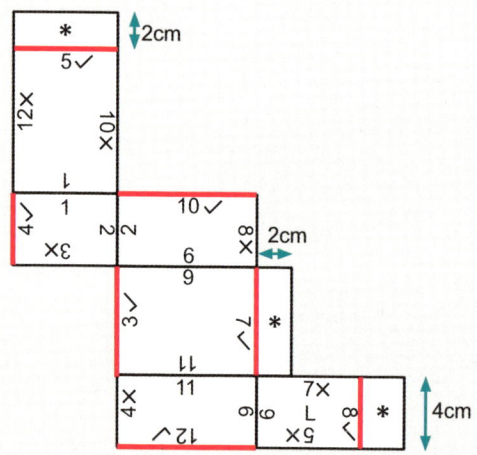

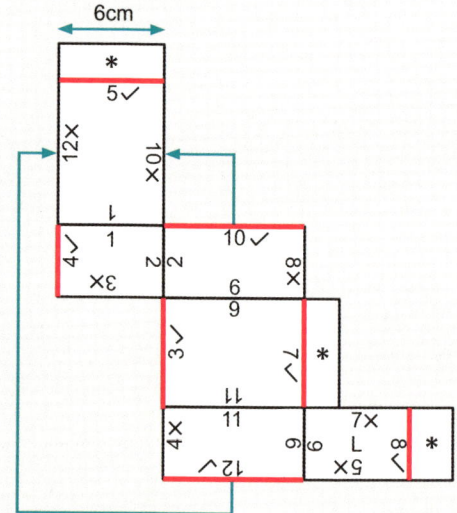

2.7.10

Cut out the 2 tabs and tape them to edges 10-tick and 12-tick.

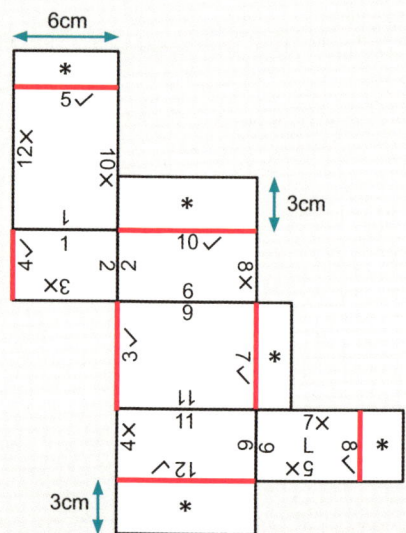

2.7.11

Finally, we can add tabs to edges 3-tick and 4-tick. They are not the same depth. The tab on 3-tick will fold under edge 3-cross. The face behind edge 3-cross is 4cm deep, so the tab on edge 3-tick will be 2cm deep (half of 4cm). The face behind edge 4-cross is 8cm deep, so the tab on edge 4-tick will be 4cm deep (half of 8cm).

2.7.12

This is the complete net.

Working out the optimum size and shape for all the tabs can be a little slow. However, if you keep folding up the box to see how it looks, you will quickly identify any problems, and a little analysis will help you to find the perfect answer. Mistakes are inevitable, but with this collage method of construction, they can be quickly rectified. Just keep pulling off any erroneous tabs and substituting them with new ones.

The size and position of every tab can be identified, so no part of the net should be designed without a justifiable reason.

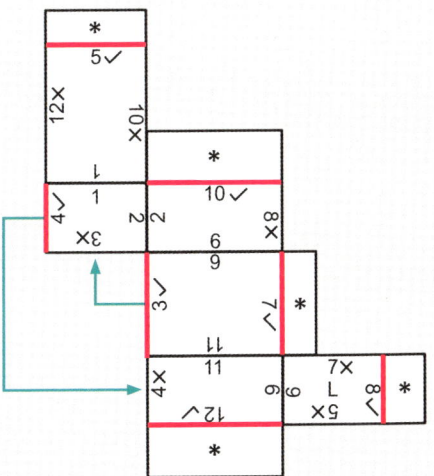

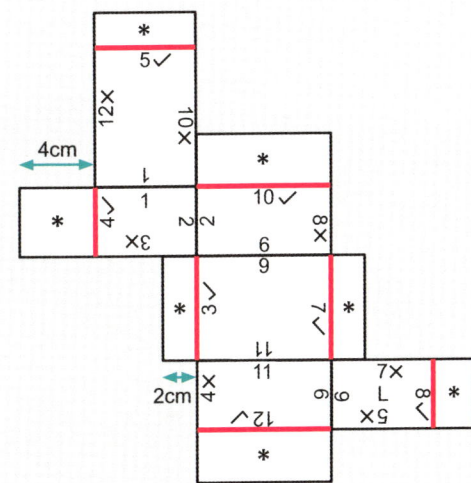

2.7.13

Here is a simplified rendering of the complete net, minus all the edge numbers. When the puzzle of the placement of the tabs and the different depths are solved, such nets acquire a logical beauty that can be very addictive!

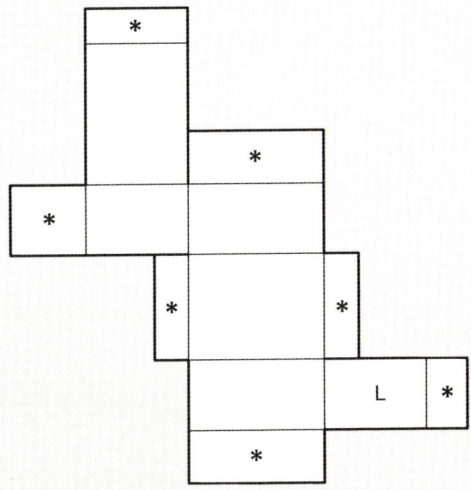

Rough-looking collage of faces and tabs, numbered and taped

2.8 Designing the Optimum Net

Staying for a while with our cuboid, of the 66 possible nets (without tabs), which nets will be better than others?

One criteria for eliminating many potential nets is to consider the area of card that they occupy. Not all nets occupy the same area. Clearly, if a net can form a cuboid with less card than another net, it is a preferred design.

Here's how to minimize your use of card.

2.8.1

Here is our 8cm x 6cm x 4cm cuboid again.

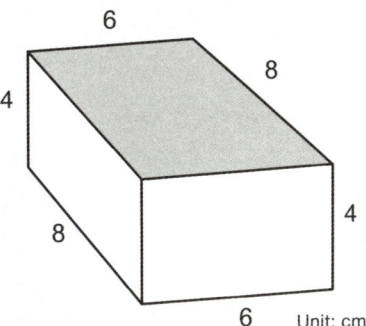

2.8.2

2 very different-looking but structurally identical nets are shown. The net on the left (without tabs) occupies an area of 560 cm^2 (28cm x 20cm). The net on the right (without tabs) occupies an area of 320 cm^2 (20cm x 16cm). This is about 43% less card than the first net!

Further, the net on the right is more compact and is thus stronger than the net on the left.

How can we design the strongest, most efficient net? Here is how.

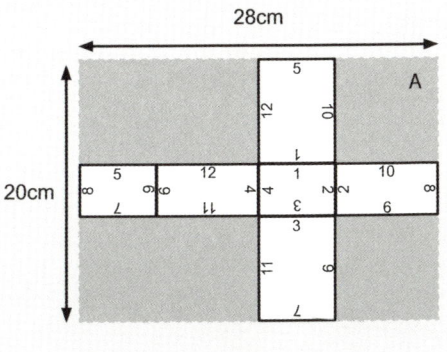

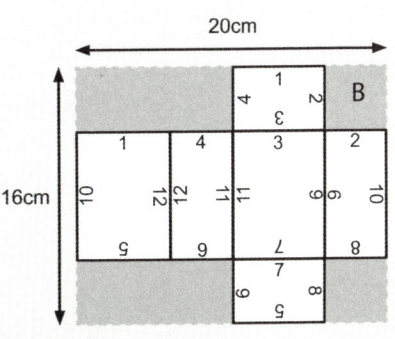

063

2.8.3

Separate the 6 numbered faces. The rule of the game is "Always join the longest edges together, whenever possible".

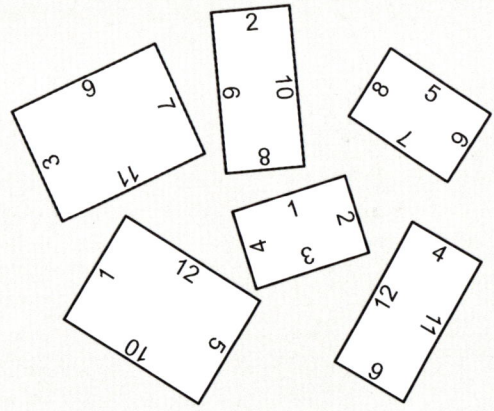

2.8.4

The longest edge is 8cm. So, we can join edges 9,9...11,11...and 12,12. This is the maximum number of longest edges that we can join.

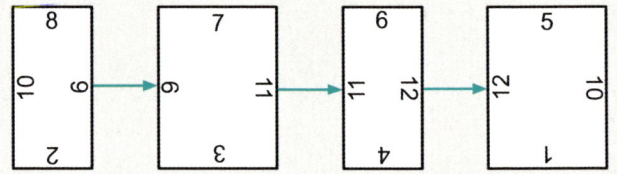

2.8.5

The next longest edge is 6cm. Edges 3, 7, 1 and 5 are 6cm long, so we can join the remaining 2 rectangles in a variety of ways, to different numbers. Here, the joining numbers are 3 and 7.

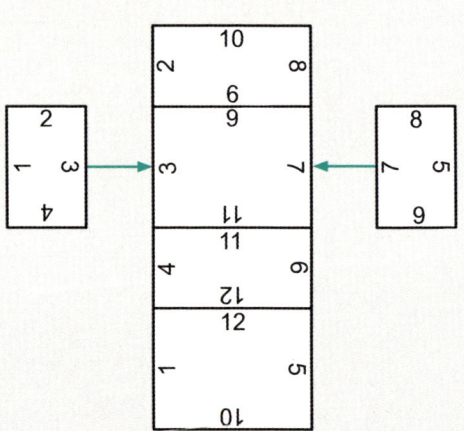

2.8.6

This is the completed net. It is the most compact and strongest net possible for this specific 8cm x 6cm x 4cm cuboid, and uses the least amount of card.

Taking the time to design your nets in this way will make you a very smart package designer! Don't be wasteful of material.

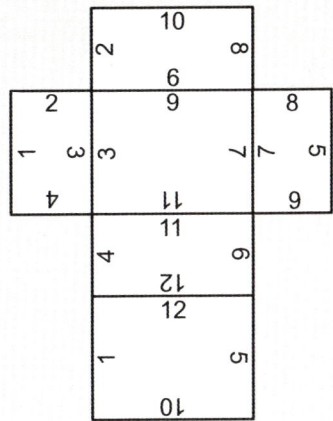

2.9 Multi-tabbing

Sometimes, your packages will have a long edge that does not lie flat and cannot be closed with glue. This is a weakness, allowing access to the interior for dirt, insects and prying hands. It must be closed.

It is possible to seal it using a Tongue Lock (see page 159).

A second way is to multi-tab. The method is simple.

2.9.1

Here is the problem edge. We will learn how to close it.

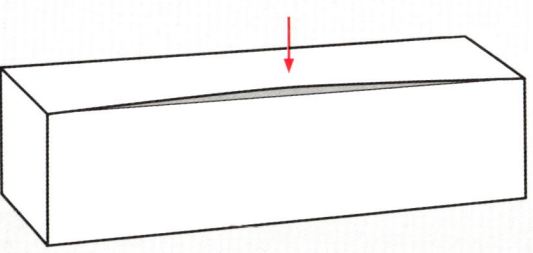

2.9.2

Here is the optimum net for the box, already tabbed.

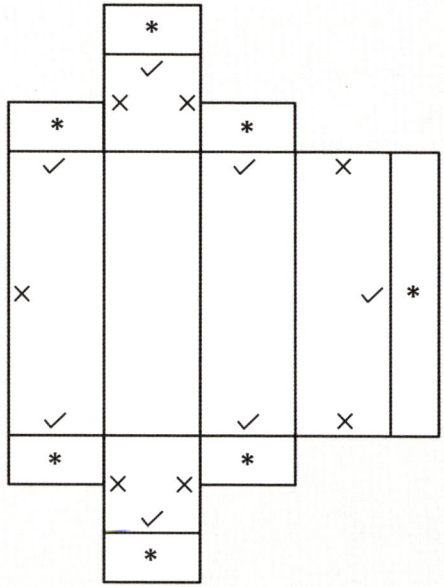

2.9.3

Divide the long edge into thirds, and draw 2 lines across the width of the net.

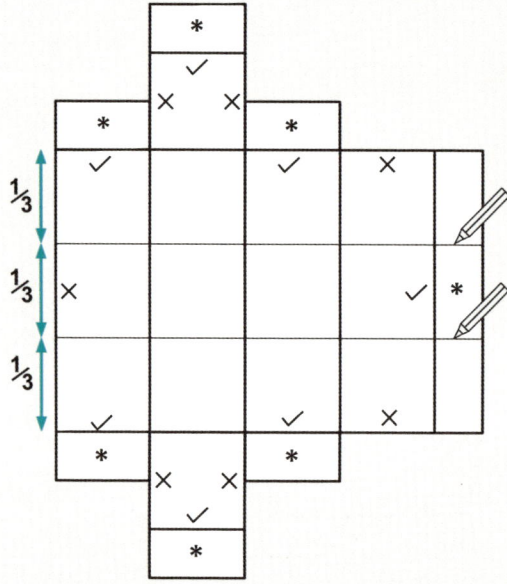

2.9.4

Remove the central third of the long tab and position it on the left-hand edge. Note that the tabs at the thirds do not have 90-degree corners. Instead, they are at about 45-degrees. This will help them to interlock with each other, knitting the edges together at those thirds. If you are assembling the box by hand, be sure to make it big enough so that your fingers can work inside the box to interlock the tabs.

Note how the tick-and-cross pattern continues around the perimeter of the box, without interruption or deviation.

It is better to divide the box into an odd number along a long edge (3, 5, 7...) rather than an even number (2, 4, 6...), because this preserves the pattern of tabs above and below the long edge.

A long edge need not be divided into precise thirds, fifths and so on. Sometimes it is advantageous to knit tabs together very close to a corner rather than away from it.

This method of multi-tabbing is not practical for mass production but works well for small numbers. It works particularly well when not designing packaging, but applying package design techniques to paper engineered gifts, exhibition displays, toys and so on, where gluing is not advised.

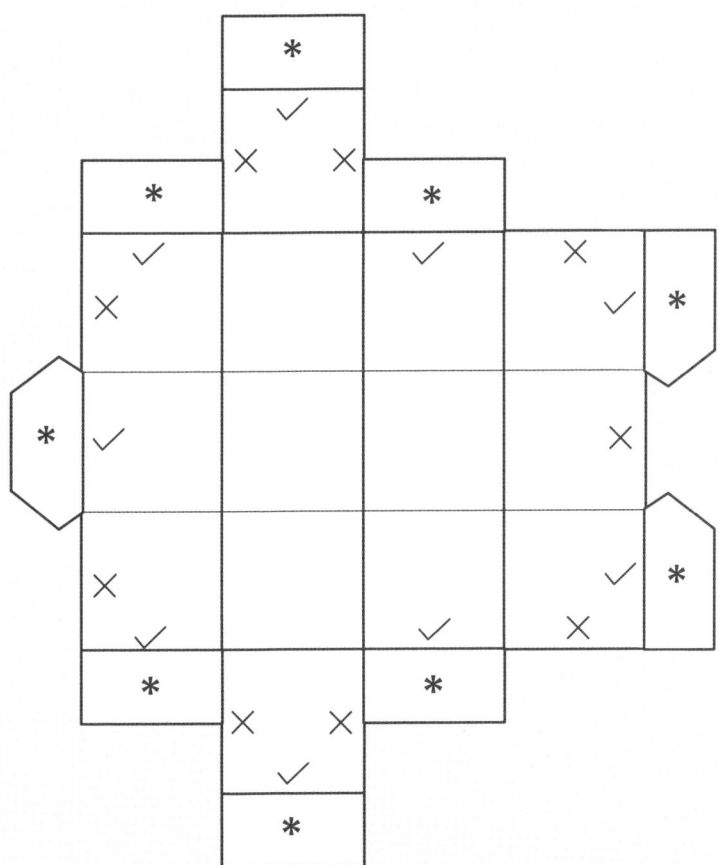

2 long and narrow cuboids, 1 with a single long tab and bending open, 1 tabbed into thirds and locked tight shut

2.10 Designing Nets with Corners that are not 90-degrees

Cube and cuboid package designs are ubiquitous because they are economic to design and easy to store and transport. But they are also oh-so dull!

If you want to add memorability to your packaging, one way is to break free from the tyranny of 90-degree corners. This section will show how the "tick...cross" tabbing system can be adapted to work with corners that are not 90-degrees.

2.10.1

This is a rhombic solid. 4 faces are squares and 2 are rhombi.

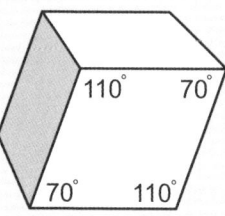

2.10.2

The net is similar to that of a cube. Note that the non-90-degree angles are 70- and 110-degrees.

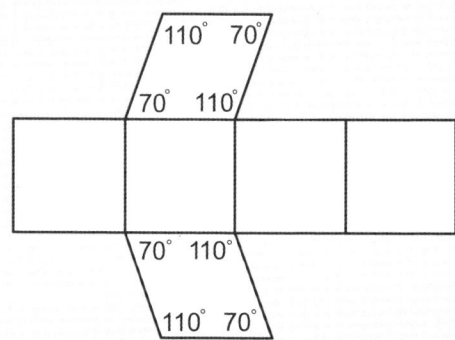

2.10.3

Using the system of numbering the edges, explained above in section 2.4, create numbers as shown (if you are making this, your numbers will be different, but will still work).

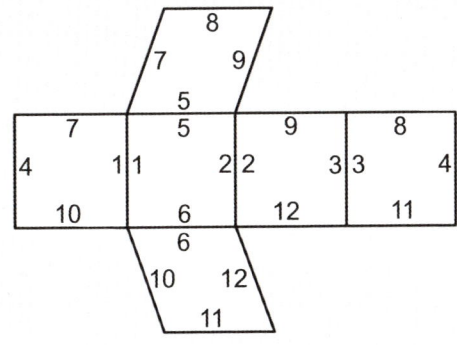

2.10.4

Starting anywhere, tick and cross the alternate edges.

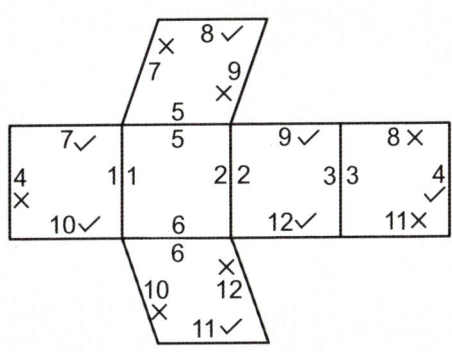

2.10.5

We are now ready to add the 7 tabs. The tab attached to edge 4-tick will slide under edge 4-cross.

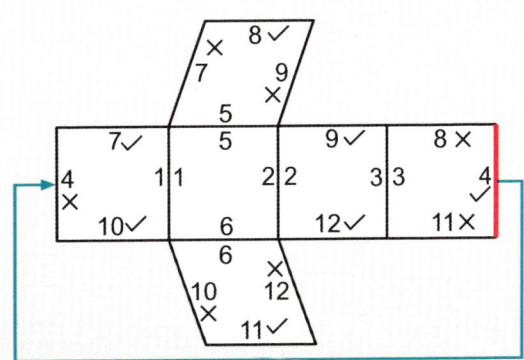

2.10.6

The gray rectangle shows the position that the tab will occupy when the box is folded into 3-D. Note that the corners at the ends of edge 4-cross are both 90-degrees, so the tab will also have corners of 90-degrees. The tab will be half the depth of the square.

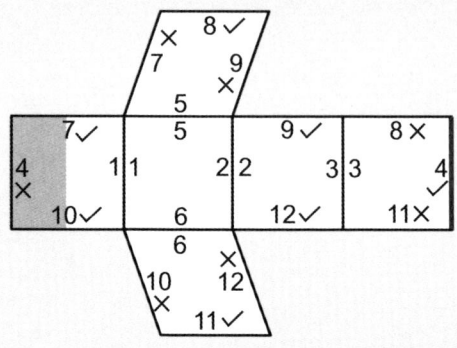

2.10.7

So, the gray tab attached to edge 4-tick will have the same shape as the gray rectangle inside edge 4-cross.

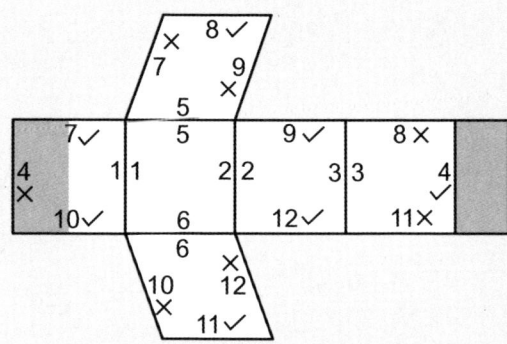

2.10.8

The tab can be cut out and attached.

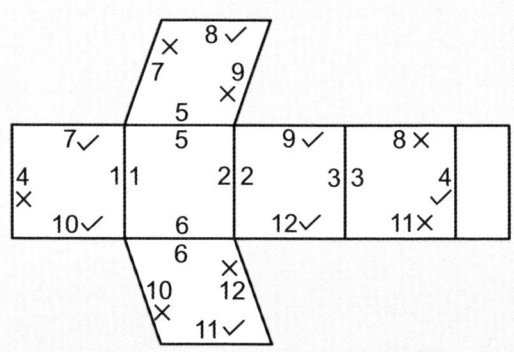

2.10.9

The edges 8-tick and 11-tick will touch edges 8-cross and 11-cross.

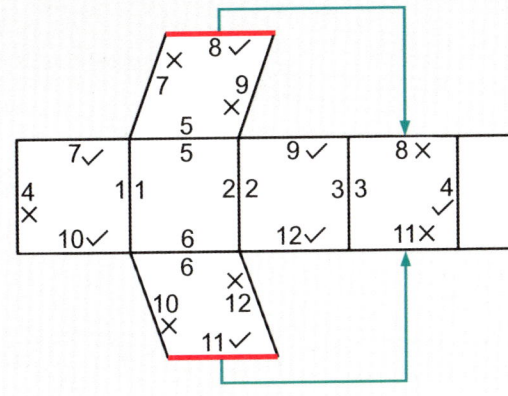

2.10.10

The corners at the ends of edges 8-cross and 11-cross are all 90-degrees, so each tab will occupy half of the face and have 90-degree corners.

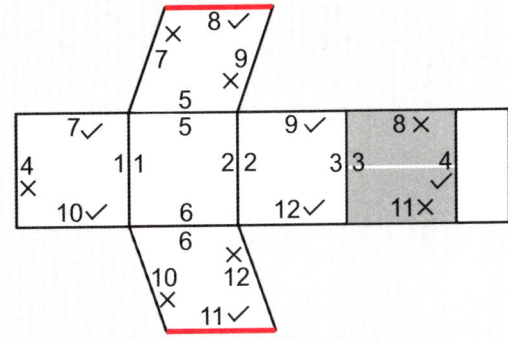

2.10.11

The tabs are here seen attached to edges 8-tick and 11-tick.

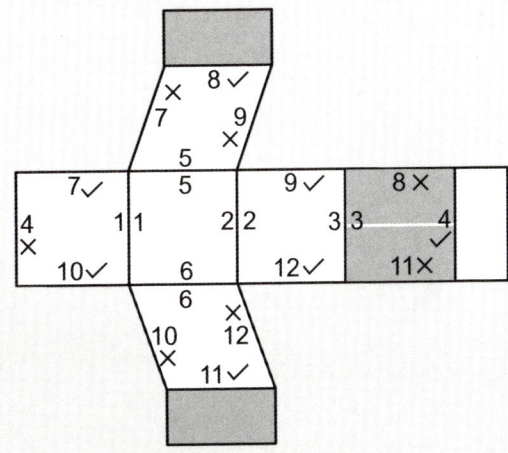

2.10.12

Cut out the tabs and attach them to edges 8-tick and 11-tick.

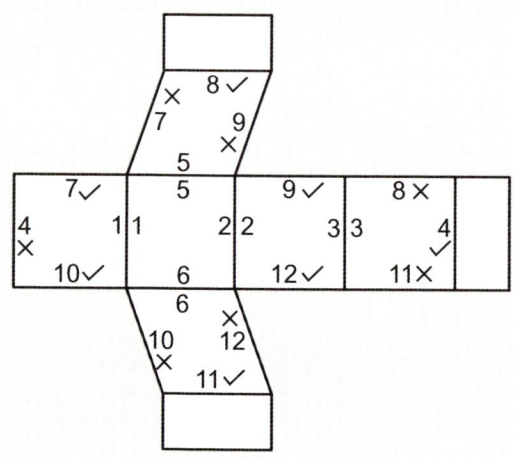

2.10.13

The remaining tabs are a little more complex, because they will not have 90-degree corners. We will begin by deducing the shapes of the tabs attached to edges 9-tick and 12-tick.

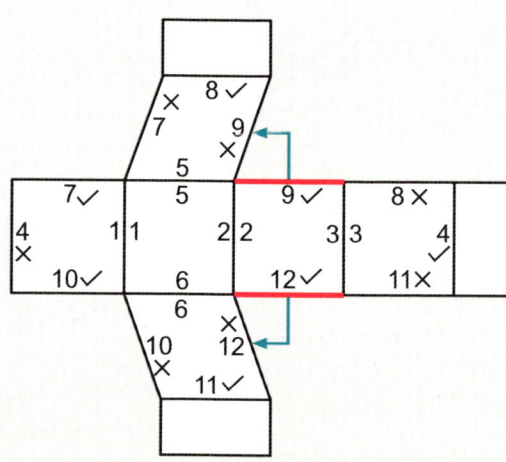

2.10.14

The corners of edges 9-cross and 12-cross are 70-degrees and 110-degrees, so the tabs will be parallelograms.

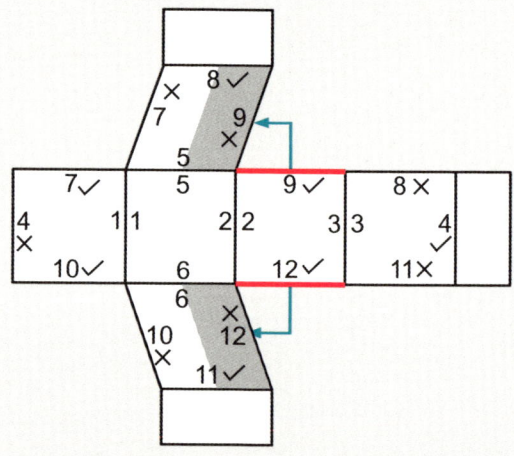

2.10.15

Further, when the net is folded into 3-D to make a box, corners A will touch corners B. It will be noticed that the angle of corners A is 70-degrees.

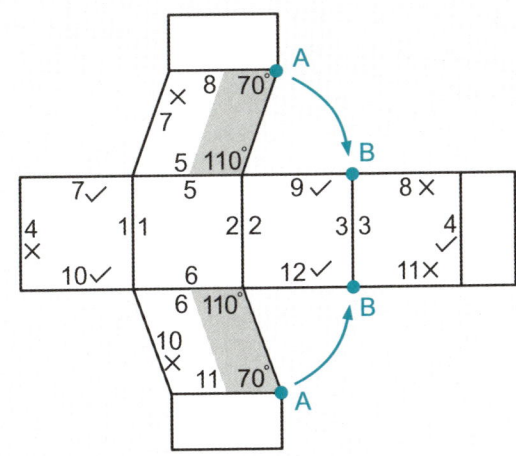

2.10.16

This means that the angle of the tab at corners B must have the same angle of 70-degrees. If one corner is 70-degrees, the other corner must be 110-degrees. However, this is impossible to construct, because the 110-degree corner overlaps the face! What should be done?

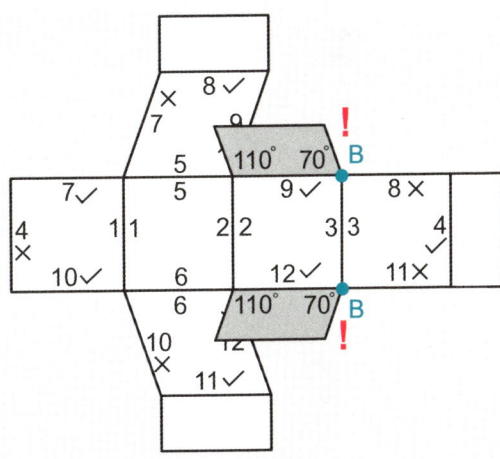

2.10.17

The answer is simply to trim the 110-degree tabs to sit flush against edges 9-cross and 12-cross. Everything will be OK. Cut out the tabs with the appropriate 70-degree angles and tape them to edges 9-tick and 12-tick.

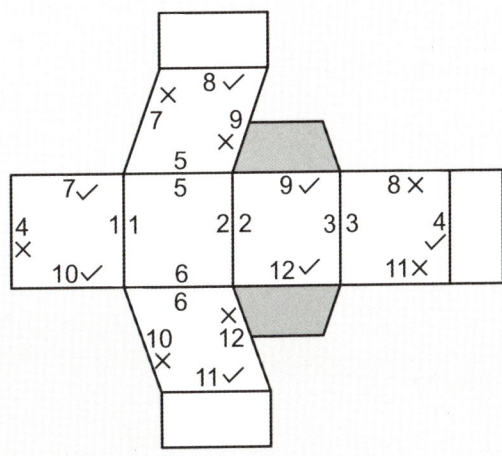

2.10.18

Finally, we can make the tabs attached to edges 7-tick and 10-tick.

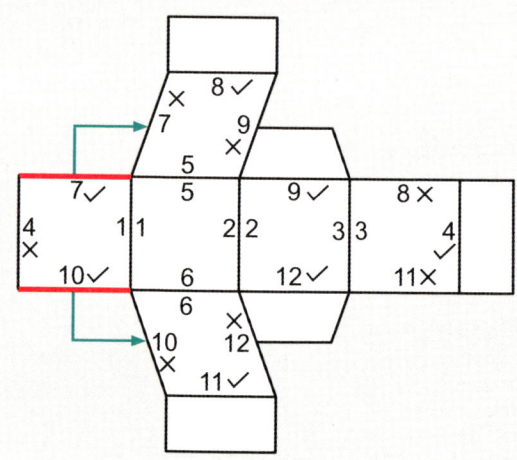

2.10.19

Notice that edges 7-cross and 10-cross have corners of 70-degrees and 110-degrees. So, the tabs attached to 7-tick and 10-tick must have the angles...but which angle will be at which end of the edges?

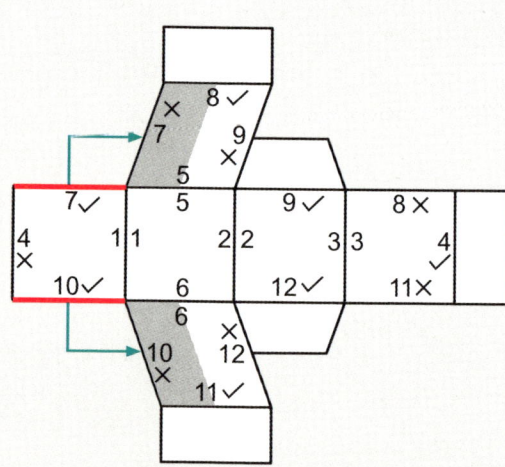

2.10.20

Notice that when the net is folded to become a 3-D box, the corners A will touch the corners B. The corners A are both 110-degrees. This means that the angle of the tabs at the corners B on 7-tick and 10-tick will also be 110-degrees.

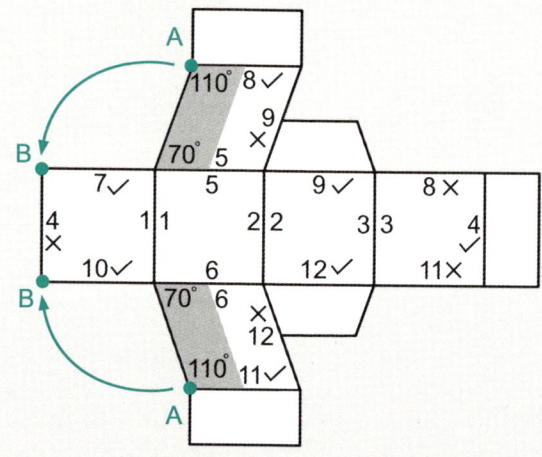

2.10.21

The tabs are attached. Note that if the angle of the tab at the corners B is 110-degrees, the angle at the other end of the 2 tabs must be 70-degrees.

Cut out 2 tabs of the correct shape and tape them to the net.

If any mistakes are found, analyze the problem, identify the mistake, remove the incorrect piece of card, make a new piece attach to the net and check that everything is correct.

This complete net will be a collage of 13 pieces of card and it will probably look rough! This is not important. Its function is to allow you to understand the configuration of the faces and the shapes of all the tabs. When you understand the structure of the net, then you can make it very carefully and it will lock perfectly.

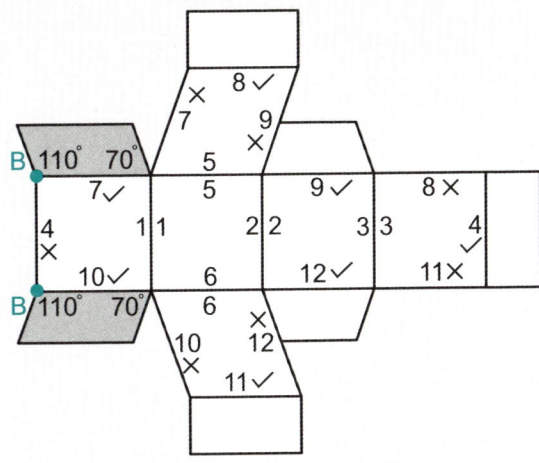

2.10.22

Here is the complete net. The placement of the faces, and the placement and shape of every tab have been carefully determined. Nothing about the net is intuitive. It is a wholly logical structure, designed according to a strict system. Any deviation from that system will weaken the design.

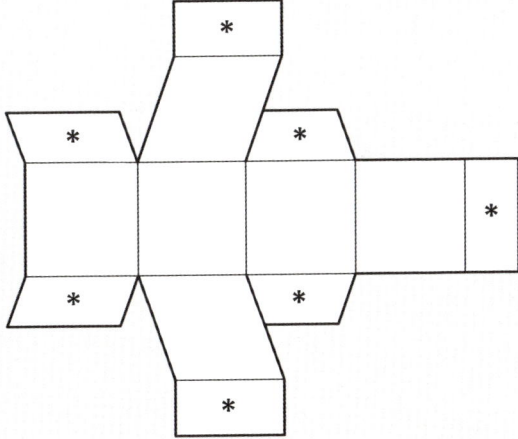

Rhombic box

2.11 How to Lock Tabs with Angles of Less than 90-degrees

The method described above works faultlessly in all cases for all nets, providing all the corners on all the faces have angles of 90-degrees or greater.

However, when the corners have angles of less than 90-degrees, the net will not lock securely. Why? Because corners such as these will generate tabs of the same angle — also less than 90-degrees — and they will not lock securely into the folded-up box. They will always fall out. It's an annoying problem!

This section will suggest several solutions to create strong and secure nets.

2.11.1 Flanges

2.11.1.1

Here is the net for a simple triangular prism. Note that the net has 2 equilateral triangles with angles of 60-degrees.

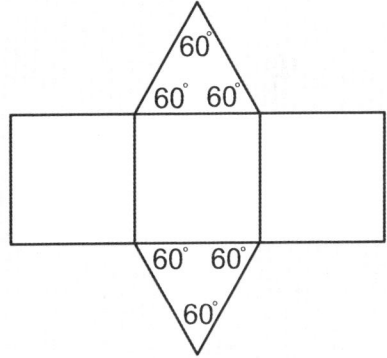

2.11.1.2

The net can be tabbed using the method described above in this chapter. Note that it has 2 tabs with angles of 60-degrees at both corners. If the net is folded up to make a prism, the prism will not lock together – the tabs will be loose.

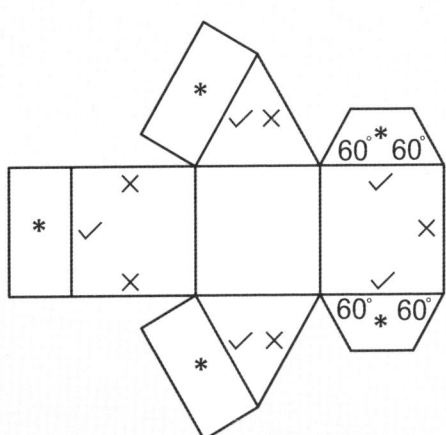

2.11.1.3

The solution is to add an extra width to the tabs, called a "flange" – here shown in gray. These flanges will help to hook the tab deeper into the net and will lock the corner tight shut.

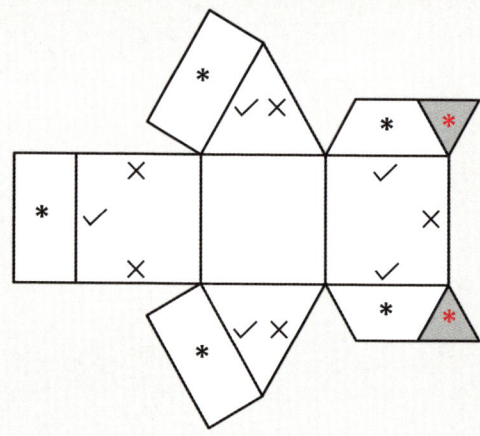

2.11.1.4

The width of the flange is more of an art than a science. In this specific case, it is sufficient to add 30-degrees. The general technique is to add as much as to the width, fold up the net, see that if it is too wide and cut back the flange a little. Fold the net again, if it is still too wide, cut back even more of the flange. There will be a point when the flange will slide easily into the net but still locks well. Remove more of it and the corner will be loose.

The general principle is simple: if you are unsure how much card to leave, leave too much and remove it little by little. Don't be tempted to guess an exact amount. Once the card is removed…it has gone!

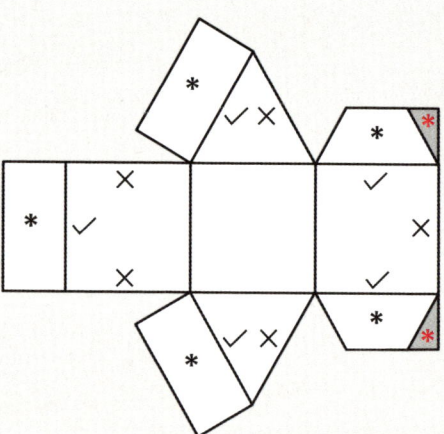

Triangular prism

2.11.2 Click Lock

The Click Lock described on this page provides another method to lock 60-degree tabs into the net. It regards the 2 triangles as lids and uses Click Locks to lock the loose lid corners tight shut. Note the turning circles on the 2 lid tabs.

The method works, but you might be dissatisfied with the small disruption caused to the smooth surface of the box by the card protruding at the Locks.

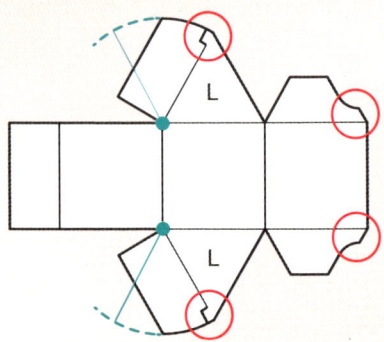

Triangular prism with Click Lock

2.11.3 Re-organize the Net to Reduce the Number of 60-degree Tabs

2.11.3.1

There are many ways that the net can be re-organized (see 2.4). One way is to surround one of the triangular faces with squares, so that it does not need to have a tab. This reduces the number of 60-degree tabs from 2 to 1. All the other tabs have conventional 90-degree corners, so locking them into the net will not be a problem.

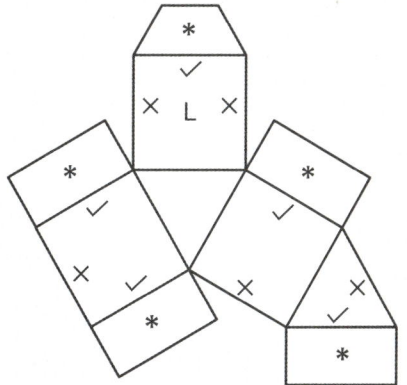

2.11.3.2

When the net is folded to 3-D, the single 60-degree tab will slide inside the triangle face.

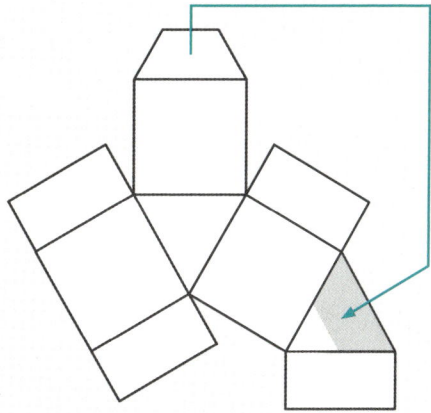

2.11.3.3

Create a semi-circular cut in the triangle face, such that the cut begins and ends at the edge of the tab.

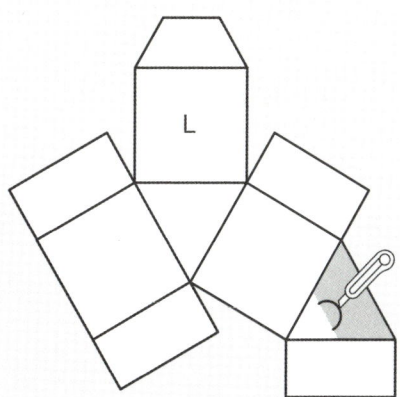

2.11.3.4

This is the completed net. In this example, the 60-degree tab will lock OUTSIDE the triangle, not inside. The semi-circular cut will hold the tab tightly to the box surface.

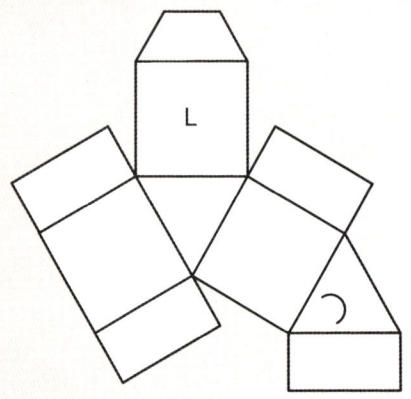

External lock

2.11.3.5

In a similar way, 2 Click Locks can be added to the corners of the 60-degree tab. The method will work, but as in 2.11.2, above, it will look a little unsightly.

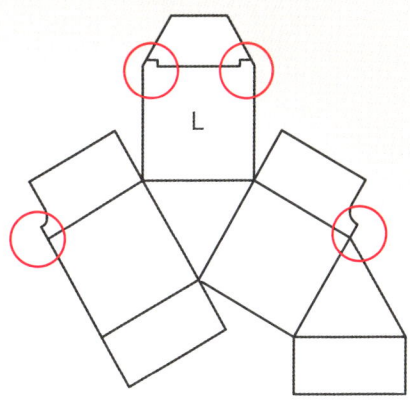

Triangular prism with Click Lock

2.11.4 Generalizing the Problem

2.11.4.1

The problem of tabs that are less than 90-degrees not locking into the net will persist, whatever the 3-D form of the box is. However, the problem can be minimized by reducing the number of corners that are less than 90-degrees at the edges of the net, requiring problematic tabs.

Look at this example. It shows a pyramid with a square base.

Drawing ① shows the 4 corners that meet at the apex of the pyramid spread far apart. When tabs are added to this net, the apex will not lock cleanly. Adding Flanges and Click Locks all the way around the 4 faces will look awful! This net, though apparently simple, is not the answer.

Drawing ② shows the apex divided into 2 parts. This is an improvement – 2 tabs less than 90-degrees are better than 4 tabs.

The best net is shown on the bottom, drawing ③. The 4 corners that meet at the apex are grouped into 1 solid corner requiring just 1 tab less than 90-degrees.

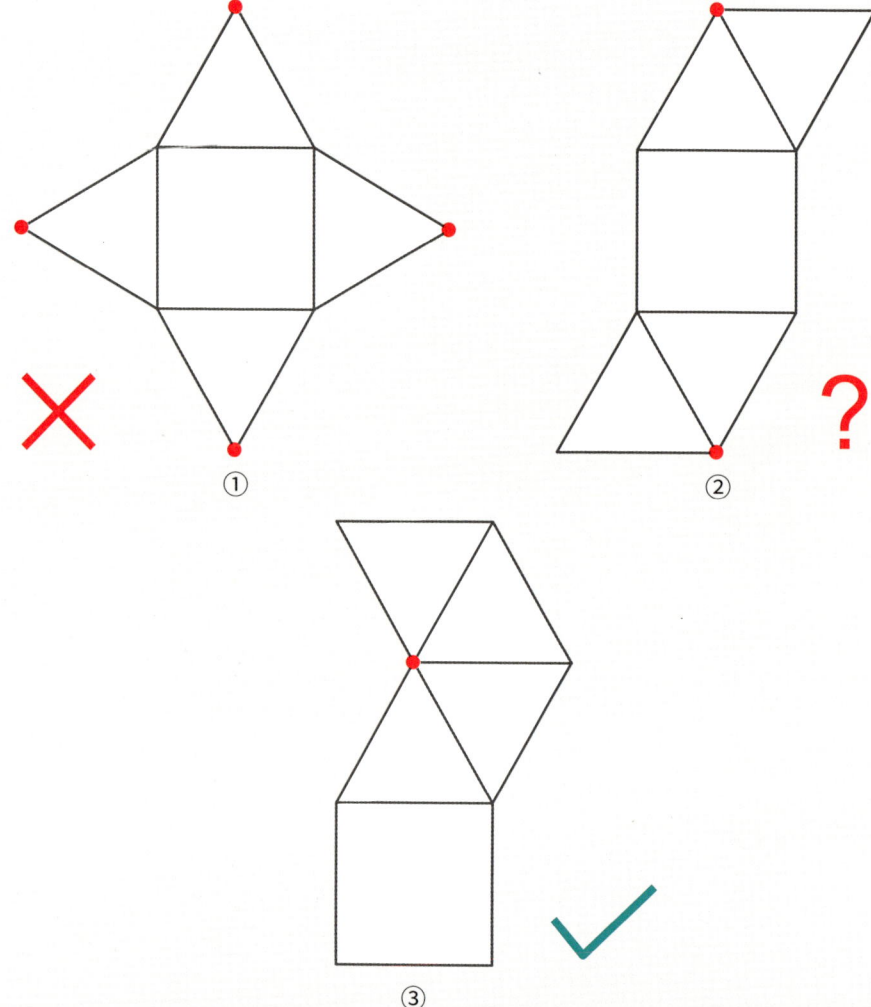

2.11.4.2

This final net can be tabbed in the conventional tick and cross method.

However, notice the difference between the 2 nets. On the left, there are 3 tabs requiring flanges. On the right, when the 4 tabs are placed instead on the other edges around the perimeter, there are only 2 flanged tabs needed! This second net will be stronger and easier to assemble than the first one.

The lessons are:

1. Always try to redesign your net to reduce the number of problematic tabs;

2. When the net has ticks and crosses on the edges, try to place the tabs on both the ticked and crossed edges, to see if one net is preferable to the other.

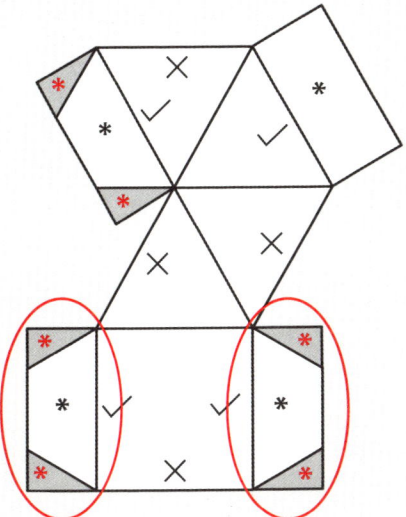

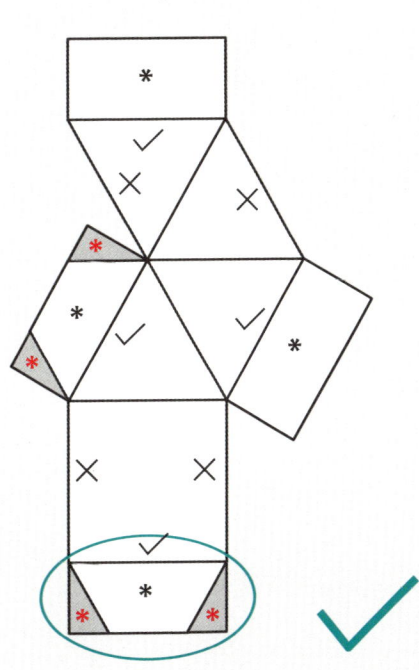

Pyramid

2.12 Designing Nets for Faces with Corners Greater than 180-degrees

There may be times when you want to design a net for a solid that has corners greater than 180-degrees. They look harmless, but dealing with these corners can be problematic, creating many strange net formations.

Nevertheless, if the principle of how to design with them is understood, your range of design possibilities increases enormously, so it has to be worth explaining.

2.12.1

In this simple solid, 2 faces (1 at the back, out of sight) have a corner of 270-degrees. This is our problematic corner.

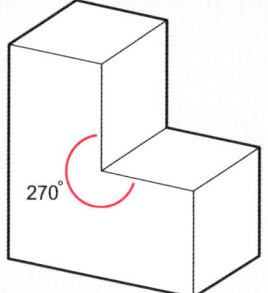

2.12.2

This is the face.

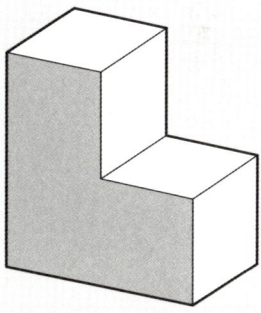

2.12.3

This is the corner.

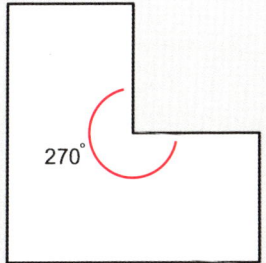

2.12.4

The solution is simply to make a cut from the corner to a nearby corner, here at the bottom left. This separates the face into 2 faces, each with corners that are all less than 180-degrees. Every corner that is greater than 180-degrees must be split in this way.

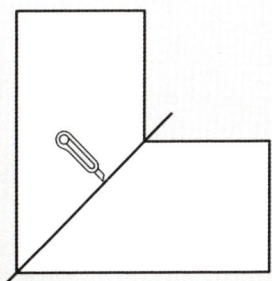

2.12.5

Note that the 270-degree corner, now split in half, is marked with two red dots.

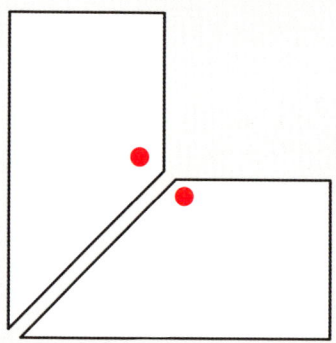

2.12.6

Here are the faces of the solid. The edges can be numbered and the pieces can be moved around until the optimum net is created.

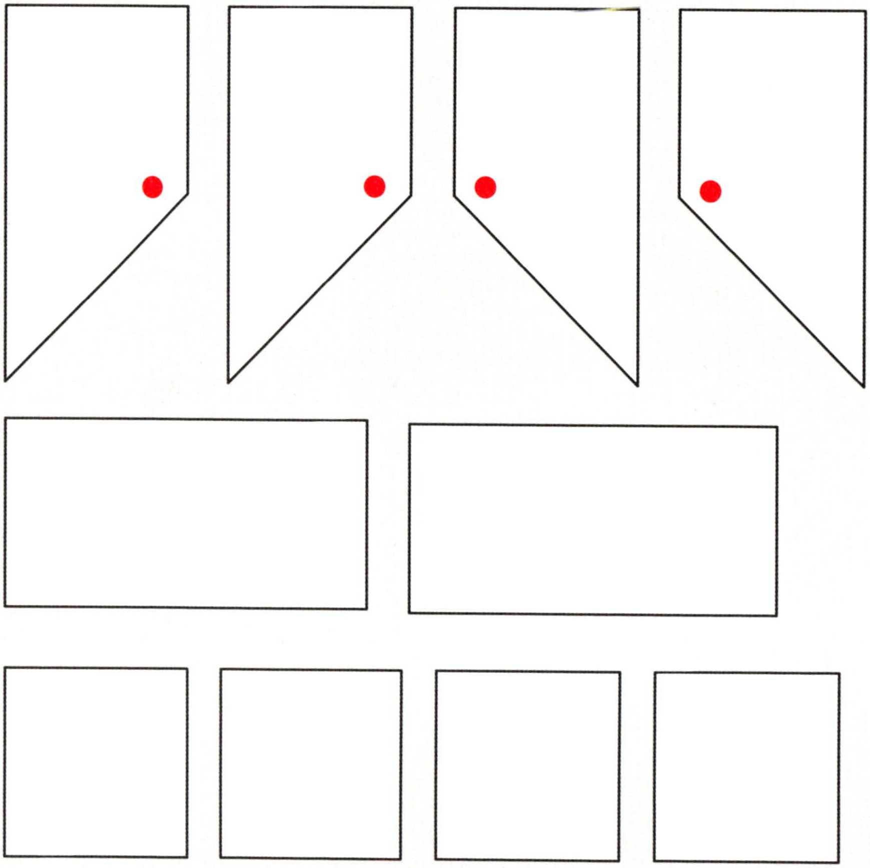

2.12.7

This would appear to be the optimum net, because where possible, the longest edges have remained connected.

However, the net is incorrect. This is because the red dots at the left edges of the net are on the perimeter, unconnected to any other face. This non-connection will severely weaken the net. All the dotted corners must be as connected as possible to other faces.

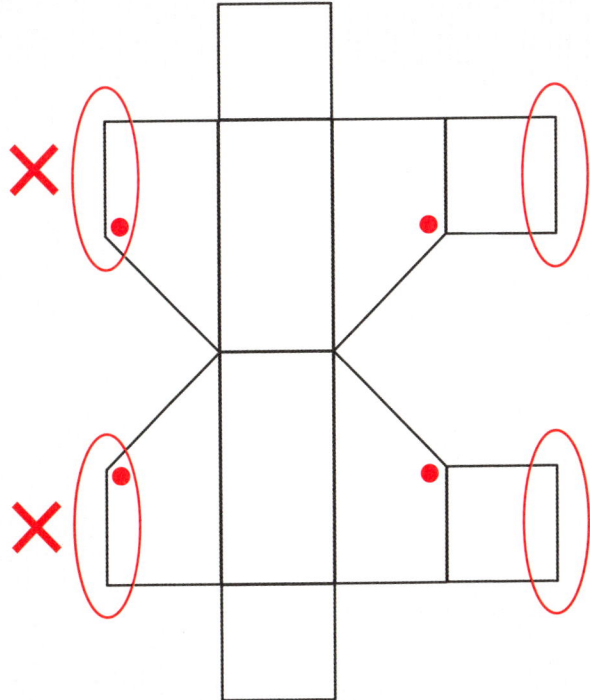

2.12.8

Here is the optimum net. Note that the long edges are not connected to each other, but more importantly, the red dot corners are all connected to another face, creating maximum strength at those corners.

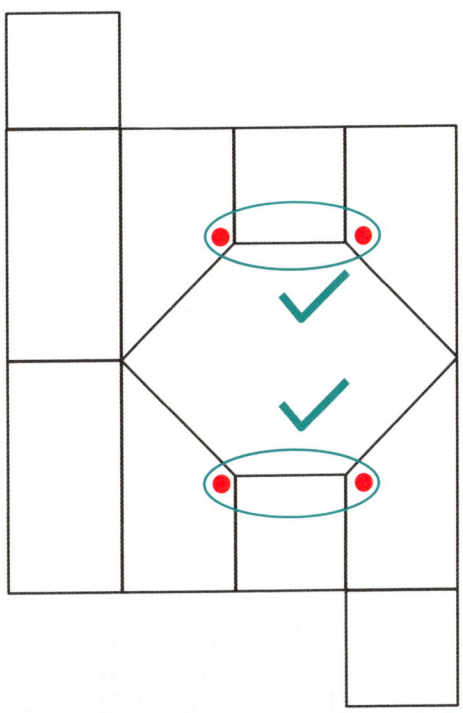

2.12.9

The net may now be ticked and crossed in the conventional way. Note that the cut made in 2.12.4 is treated as an edge.

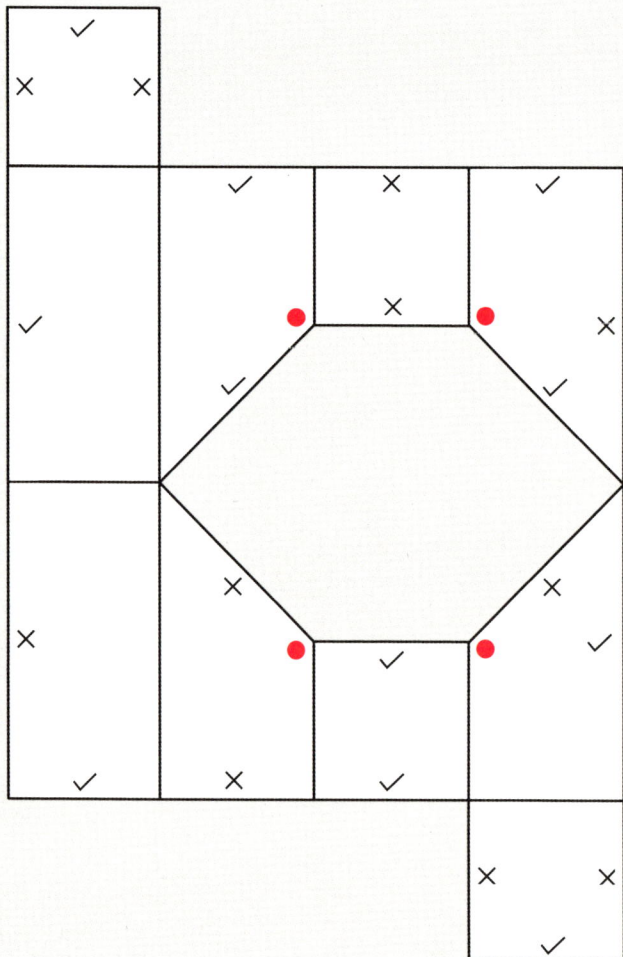

2.12.10

Attach tabs to the ticked edges, including to the cut edges.

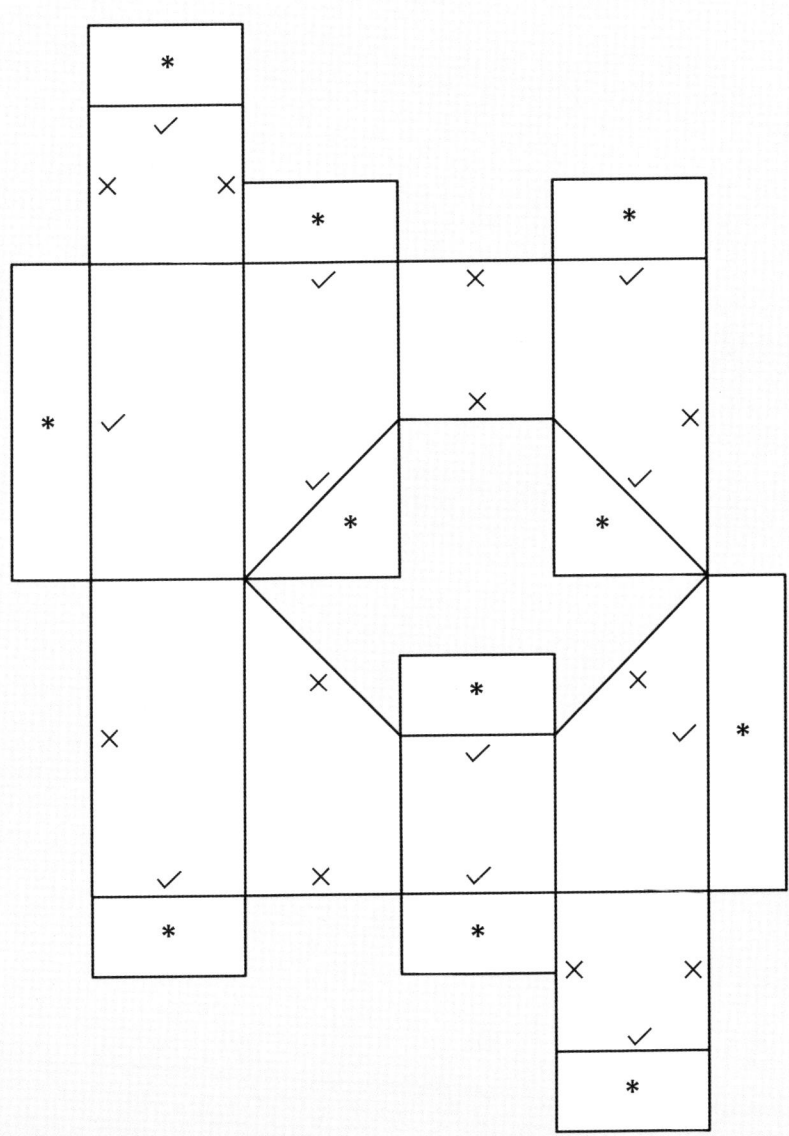

2.12.11

Here is the peculiarity of tabbing this net. The folds that should be made connecting the 2 triangular tabs to the cut edges are not to be folded! The cut edges stay flat, transitioning invisibly from face to tab. It looks very strange, but it is correct.

The tab of the square face on the bottom half of the net is also left unfolded.

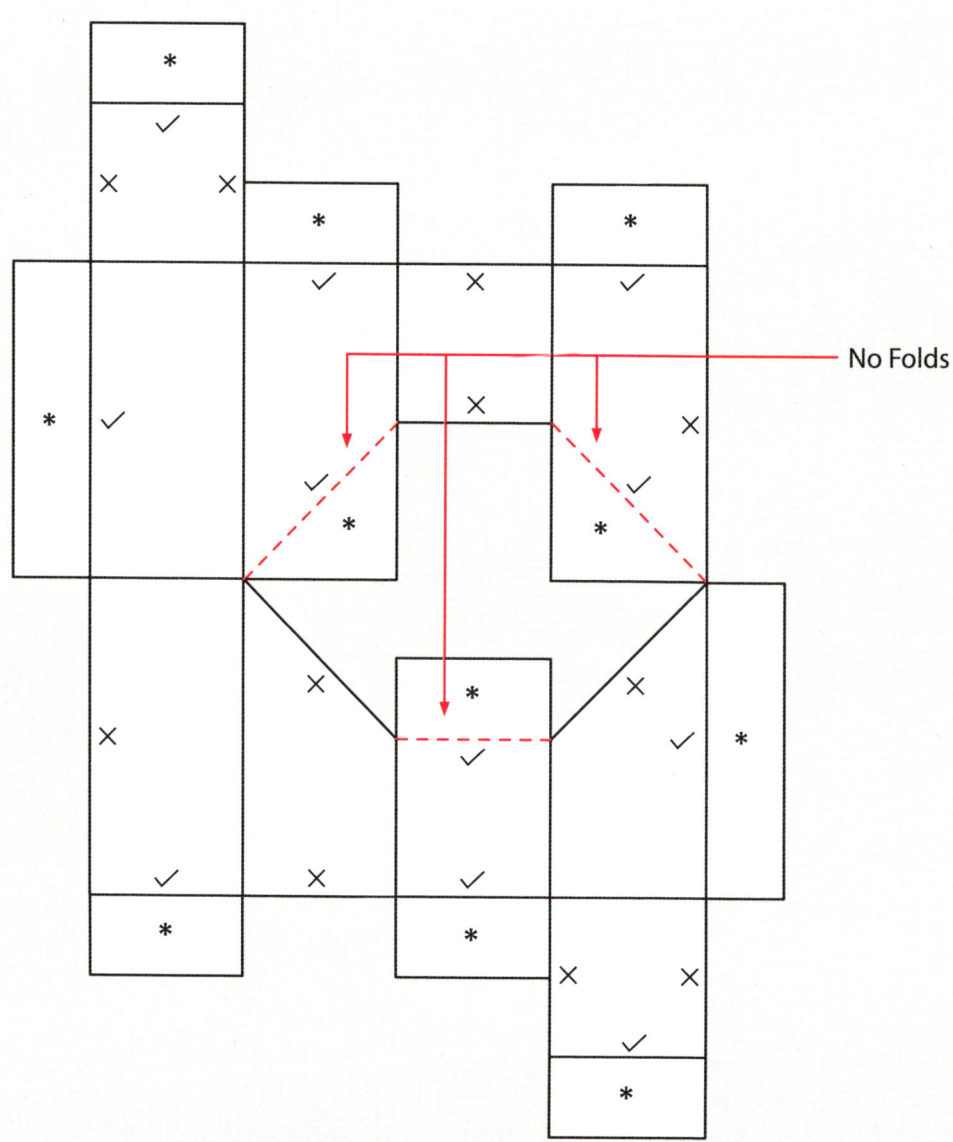

2.12.12

Finally, we need to add 2 large flanges to help hooking the triangular tabs into the body of the net, so they lock solidly. They are also to be left unfolded.

In truth, adding these strange, unfolded tabs is very difficult to anticipate and it should be done step by step, tab by tab, on a rough model. Make as many mistakes as you need to make taping and gluing tabs into different positions. When you are sure the net is as sensible as you can make it, that is the time to make a net carefully and to see if everything works OK. There is simply no substitute for experimentation and learning from mistakes.

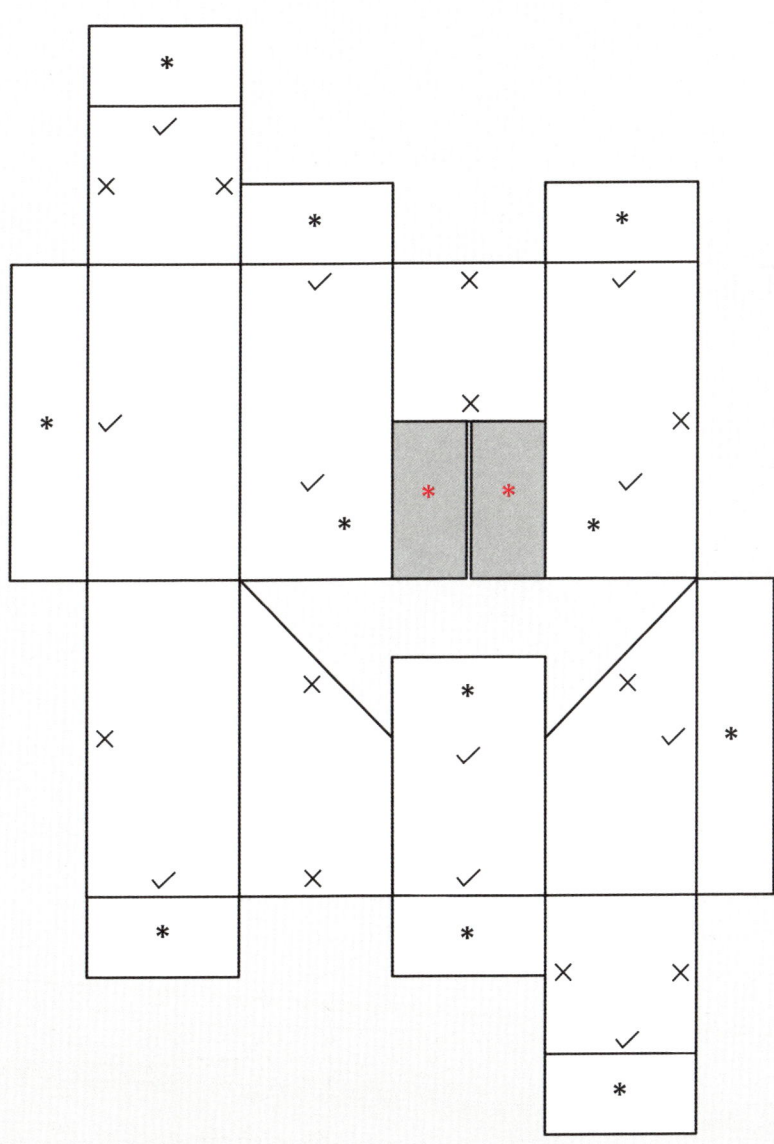

2.12.13

This is the completed net. It may look a little unorthodox, but it will work!

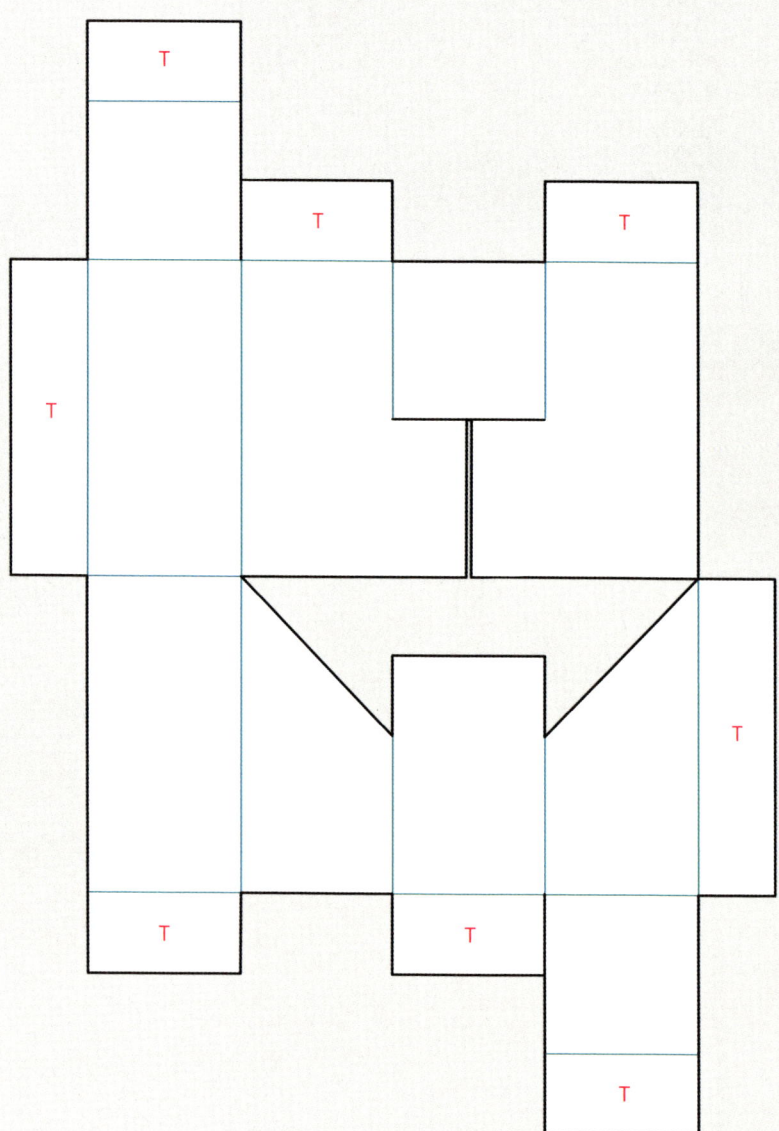

L-shaped box

03

CUBE DISTORTIONS

P101   **3.1 A Catalogue of Distortion Themes**
P136   **3.2 Strategies for Creating Multi-packs**

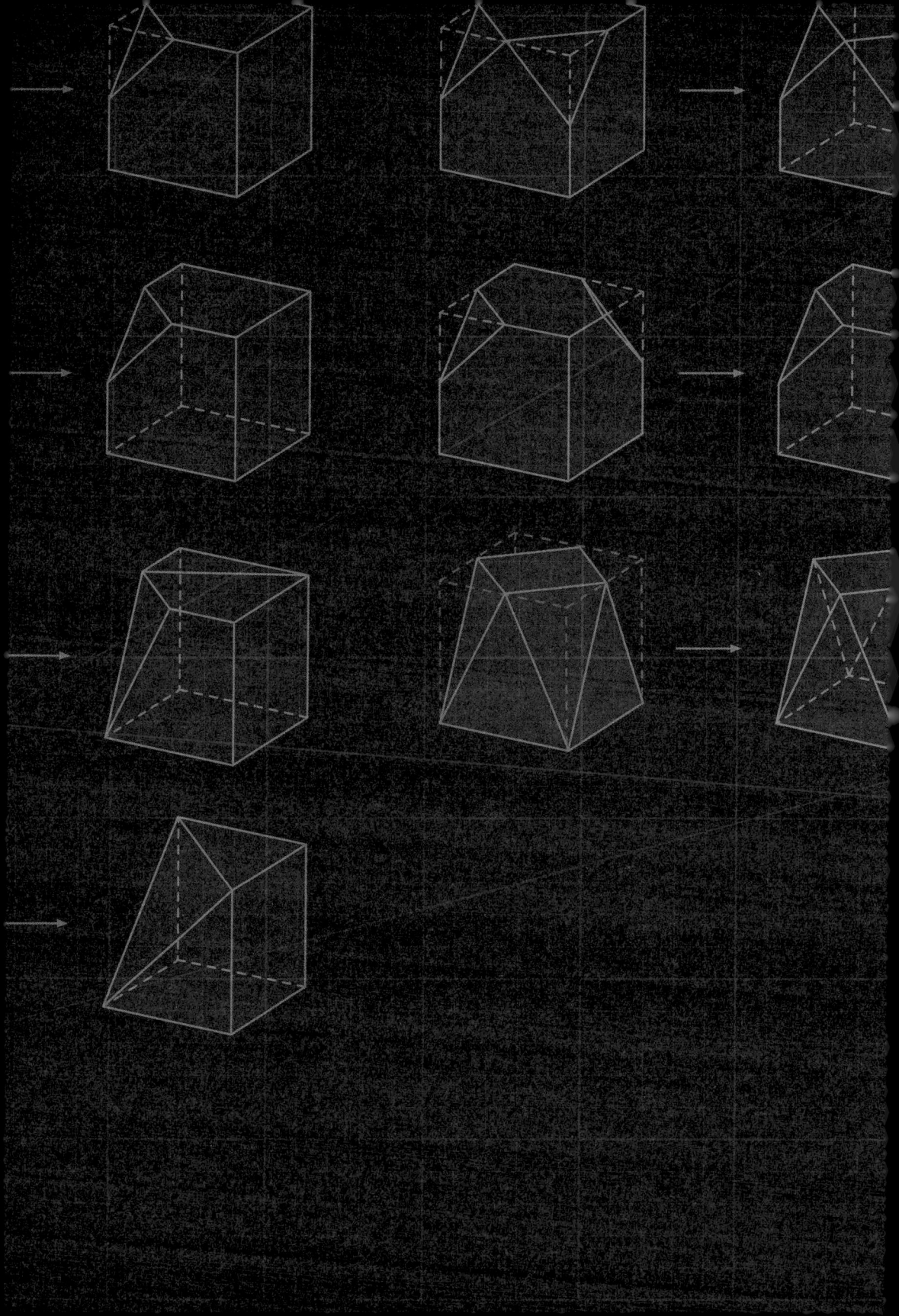

The previous chapter taught methods for constructing strong nets. It showed how to construct them for cubes and cuboids, and also how to construct tabs for angles other than 90-degrees.

However, what it did not show was how to create an original form. Cubes, cuboids and simple 3-D forms such as pyramids and prisms are very practical packaging forms, but if you are seeking to develop unique and innovative forms, how and where should you start?

Through the many packaging projects that I've run for professionals and students, I have evolved a simple but effective method to create original forms. It is based on the simple idea of taking the ubiquitous cube – a form that is so well known that it is almost a cliché – and distorting it in some way. It can be subjected to a series of processes that morph it from its familiar shape to one that is related, but still unusual, perhaps unique.

These processes involve imagining the cube is made either from a solid substance, such as wood, or from a stretchable material, such as solid rubber.

A wooden cube can be taken to a sanding machine to have an edge shaved back, or a corner blunted, or a face shaved at an angle. The shavings can be done in combination, or the same reduction can be repeated on opposite or adjacent edges, corners or faces. This will always result in a form smaller than the original cube, but one that is clearly derived from it.

A rubber cube can be stretched, compressed and twisted in different ways. For example, opposite corners, edges or faces can be stretched apart, or be compressed. There are different ways to twist a cube, using different axes of symmetry.

Finally, an unrelated technique is to substitute curved or double-curved edges for straight edges. They can radically alter the look of even the simplest form.

These simple manipulations will quickly create unusual 3-D forms. They are generally simple to make and often look beautiful. The net construction methods explained in the previous chapter ensure they will all be strong and practical.

This chapter will introduce the main manipulations and suggest variations for others. At the end of the chapter, an explanation of how a cube can be dissected into identical multi-pack forms, based on lines of symmetry through faces, edges, or corners, will further explore the concept of creating novel forms from the familiar cube.

This may seem to be a dry subject, or one which needs an extraordinary 3-D imagination or geometric know-how. But you would be wrong! I have seen a great many nervous beginners distort a cube in the simplest ways, then make a second-level piece and then a third-level... by which time the forms not only become unique, but remain practical.

The creative potential of this method is almost without limit. Please read the chapter carefully and make as many examples as possible. There is simply no substitute for making a form and observing its characteristics by turning it around and around in your hands.

3.1 A Catalogue of Distortion Themes

3.1.1 Angled Face

3.1.1.1

The distortion is in the edge length "X" and the angles "a" and "b". After these have been drawn and calculated, the edge length "X1" and the angles "a1" and "b1" can be added to the net. Thus, the lower half of the net cannot be completed until the top half has been drawn, and all the edge lengths and angles calculated.

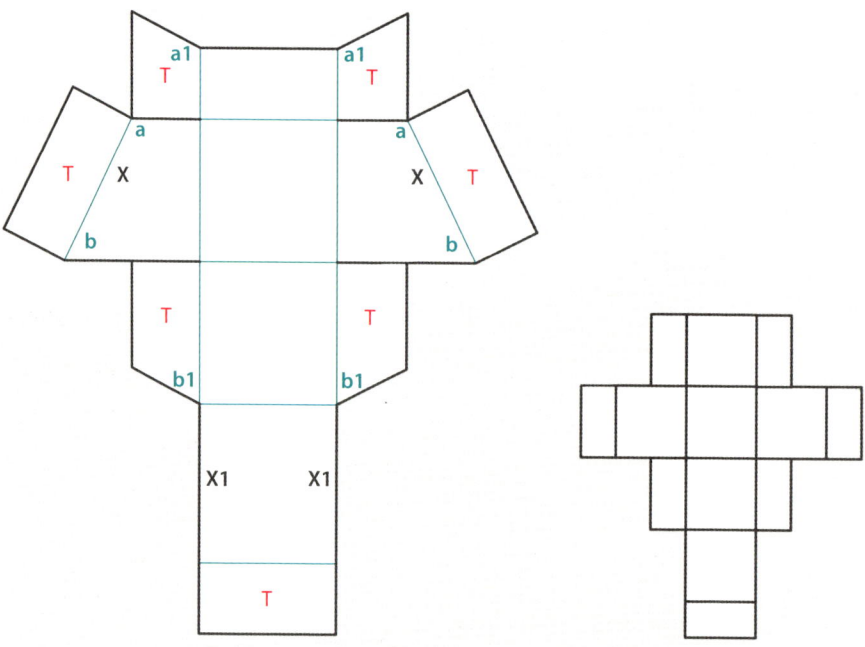

3.1.1.2

Here is the distortion. The simple cube on the left has its top face cut at an angle. The front-left face is reduced to half of its original height, but the cut could be made at another angle.

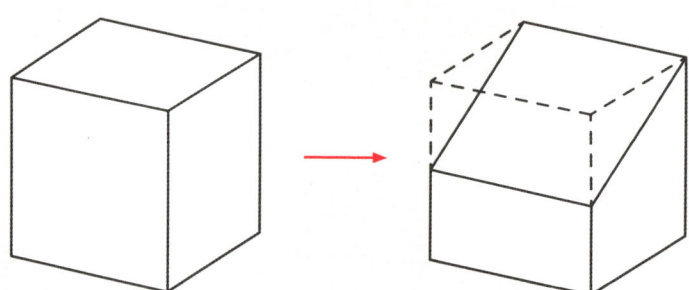

3.1.1.3

Variations.

There are 6 faces on the cube, so any face can be cut back at an angle.

When 2 faces are cut back, the relationship between those faces becomes crucial. Further, it is not only the choice of faces that is important, but in which direction each face is cut. Since there are always 4 directions to choose from. Thus, there are very many available possibilities, each resulting in a unique 3-D form with a unique net. Introduce a third and a fourth cut and the number of possible variations becomes almost incalculable! Factor in distorting a cuboid and changing the angle/s of the cut/s and we have a near-infinite set of forms, just from this simple idea.

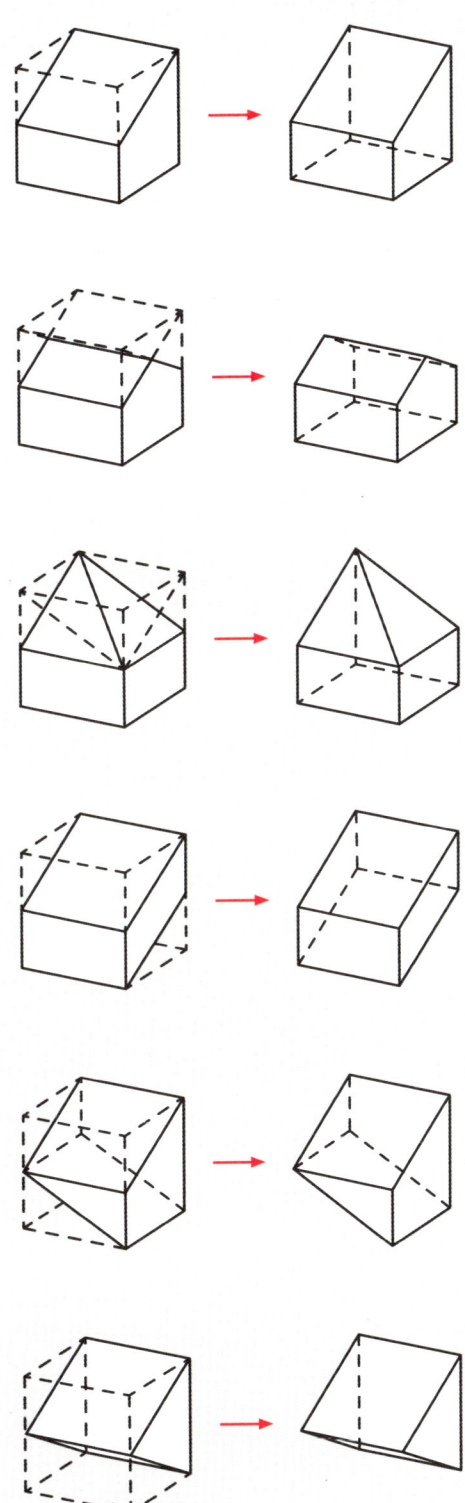

3.1.2 Shaved Edge

3.1.2.1

Here, 1 edge is shaved back to create a seventh face, shown by the additional edges at "X". Note the angles at "a". When "X" and "a" have been measured, edges "X1" and angles "a1" can be added to the net.

Note how the tabs near "a1" have been cut back. The long sharp points are unnecessary and unwieldy, so it is best to remove them.

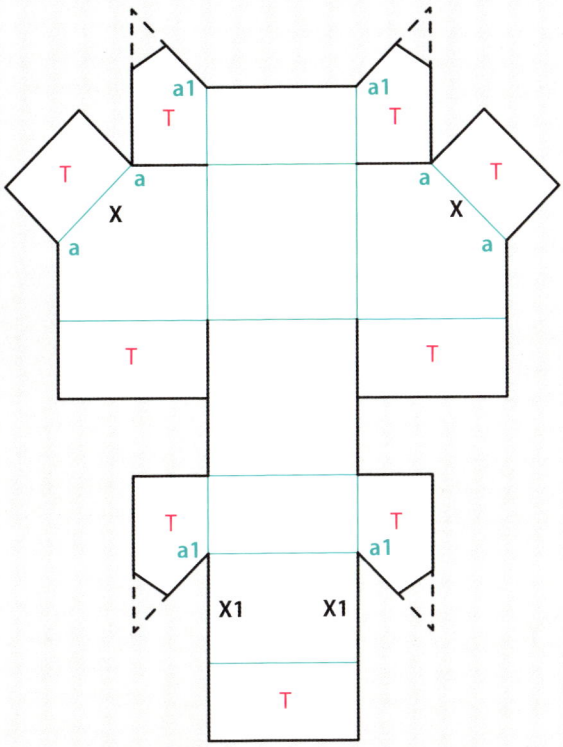

3.1.2.2

Half of the top face has been removed, and similarly, half of the front-left face has also been removed. The result is to shave the missing edge back and back, until it creates a rectangle.

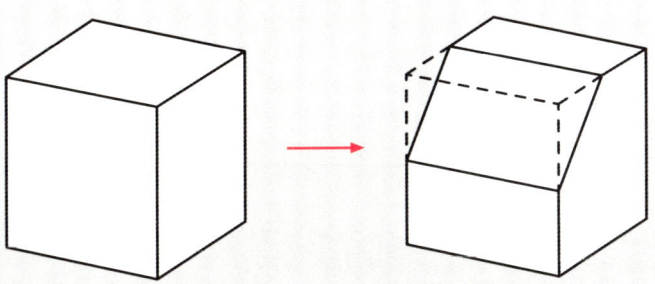

105

3.1.2.3

There are 12 edges on a cube, so there are a great many different ways in which 2 edges can be shaved. When 3 or more shaved edges are created, the possibilities approach infinity!

Note how many of these sample forms can be set down on different faces to create a very wide variety of attractive 3-D forms which appear to have little relationship with a cube. Packaging approaches sculpture!

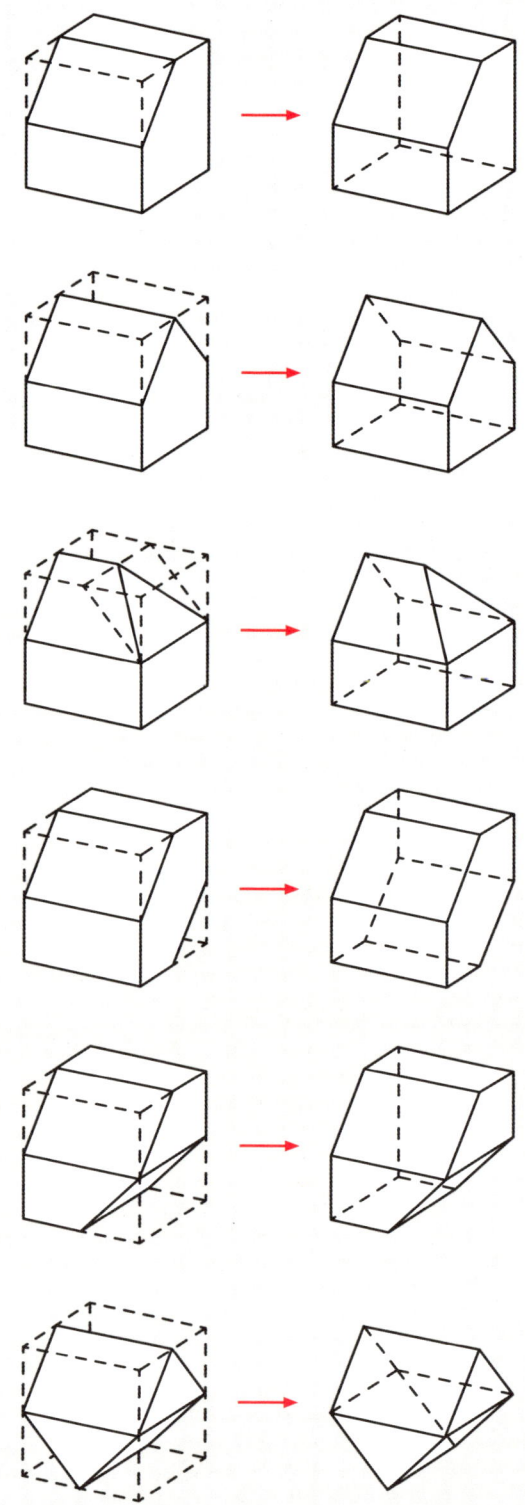

3.1.3 Shaved Corner

3.1.3.1

A corner is shaved back to create an extra face in the shape of an equilateral triangle. The creation of the triangle creates several different angles on the faces of the cube, and consequently, also on the tabs. Calculating the angles at "a" and "b" will give the angles at "a1" and "b1".

Note that the net is asymmetric. This is to create extra space around the perimeter of the net for the tabs, especially at the bottom of the net. A conventional net would have little space for generously proportioned tabs.

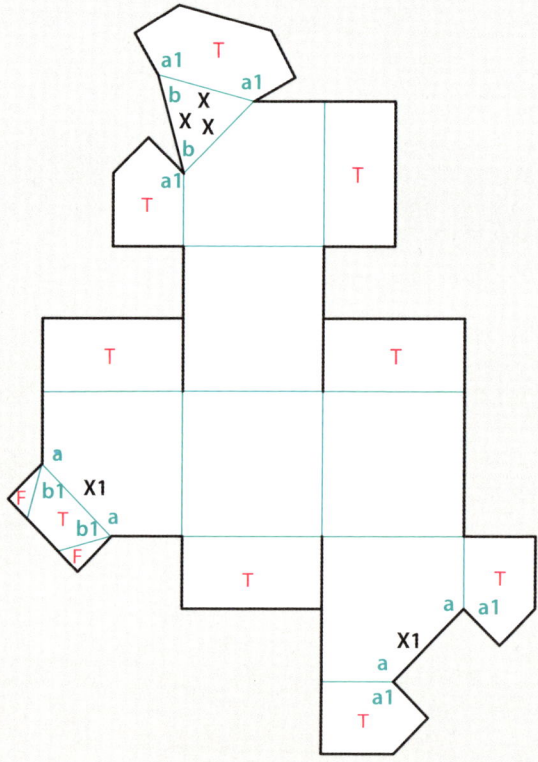

3.1.3.2

The corners of the triangle are all at the midpoints of the 3 edges on the cube. The appearance of a triangle on a cube is surprising! Somehow it seems like a simple cube with angles of only 90-degrees could never generate such a different polygon so easily.

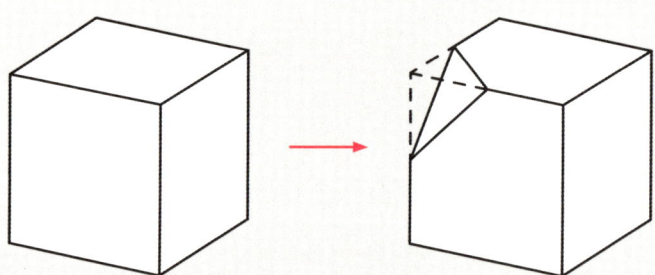

3.1.3.3

The 8 corners on a cube can each be shaved back. The size and shape of each triangle is entirely the choice of the designer, and they can be equilateral, isosceles, scalene, large or small. Triangles can be created and combined in very many interesting ways.

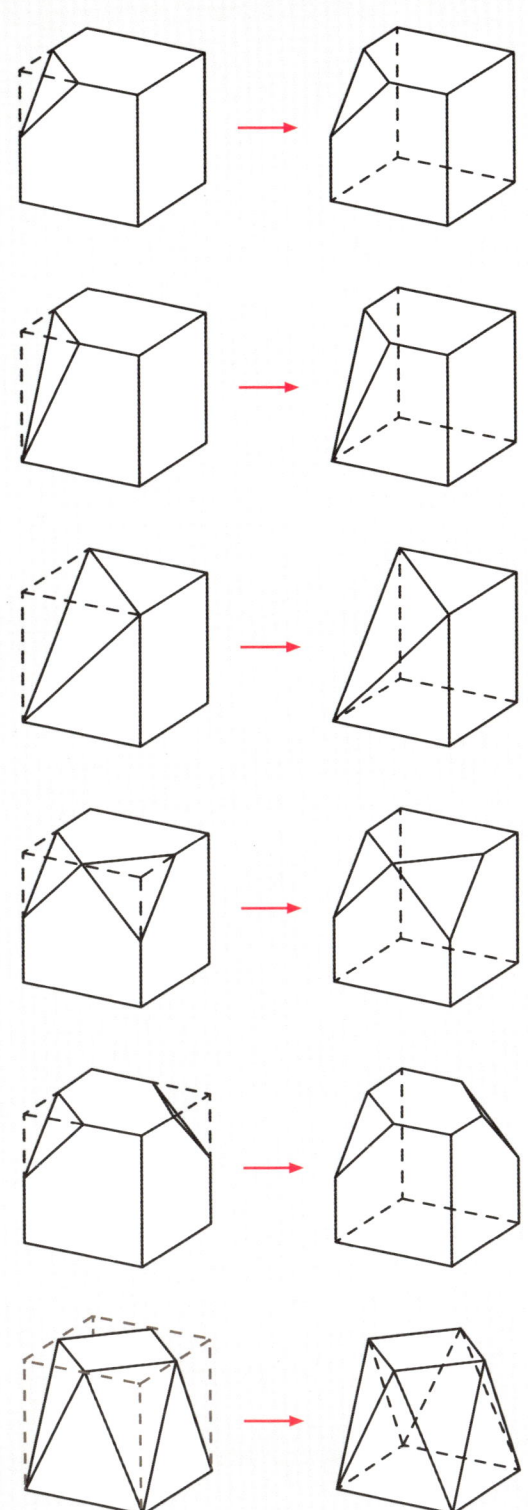

3.1.4 Inverted Corner

3.1.4.1

This is one of the simplest but most fun distortions to make. It is simply the net for a basic cube, but with the addition of extra folds around 1 or more of the corners that will invert that corner into the cube. It is similar to the Shaved Corner seen in 3.1.3, but instead of a flat triangular face being created, the corner is inverted.

It is best done on the net around a corner where 3 faces meet and where a tab fills the remaining 90-degrees, thus creating a complete 360-degrees of card.

This creation of a negative corner means that the positive corner of another cube (or other forms) will nest into the inversion, mating the corners solidly together. The implications for non-packaging uses, such as displays, are huge, as this mating technique means that an infinite number of sculptural forms can be created.

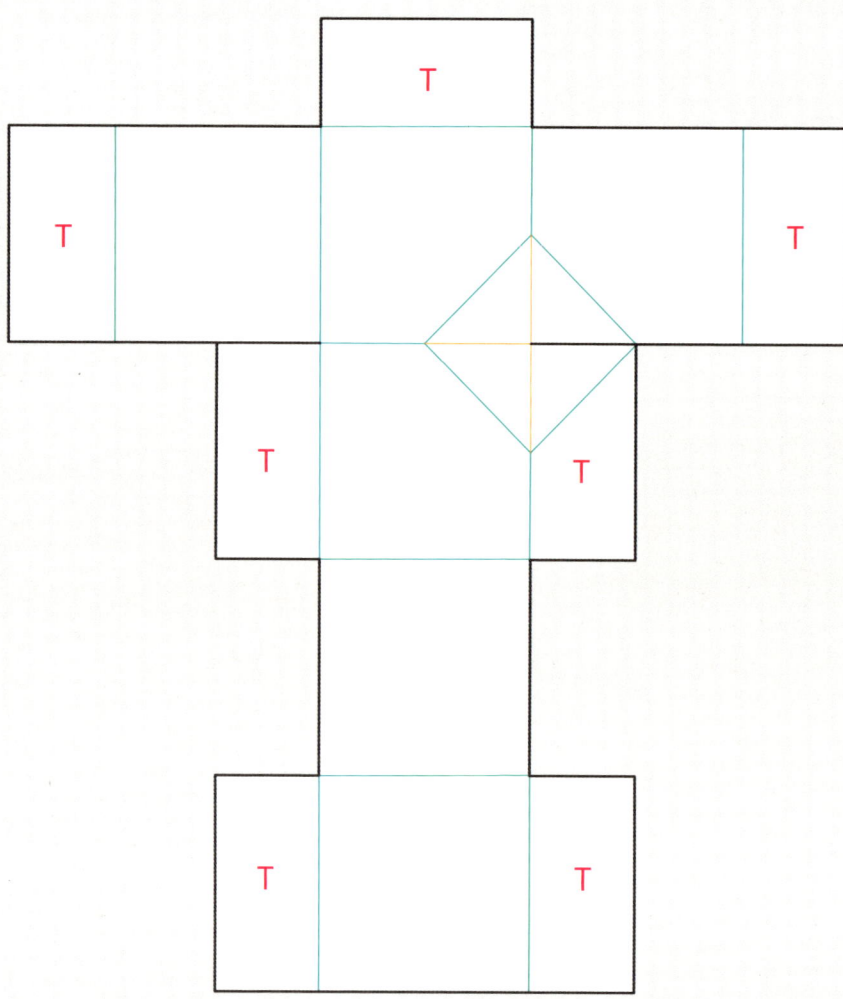

3.1.4.2

The technique is simply to invert the corner. It can even be done on the corner of a manufactured box, using nothing more sophisticated than sheer brute force against the corner!

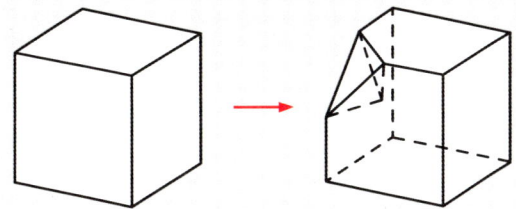

3.1.4.3

There are many possible variations of size and shape. It is even possible to split the technique onto separate faces and tabs around the net, though this may somewhat weaken the final structure. The ultimate example is to invert all 8 corners. This is quite a challenge, but the result is definitely worth it!

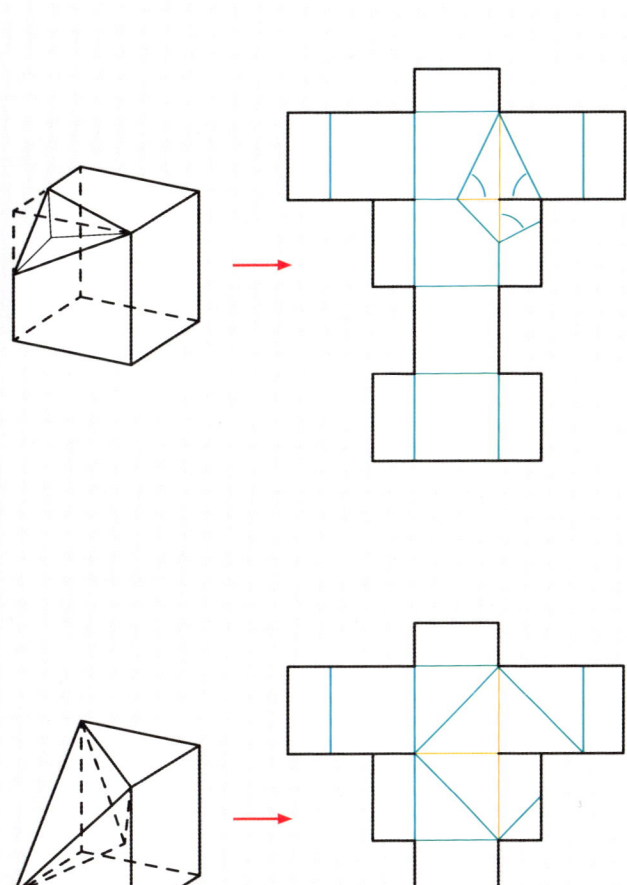

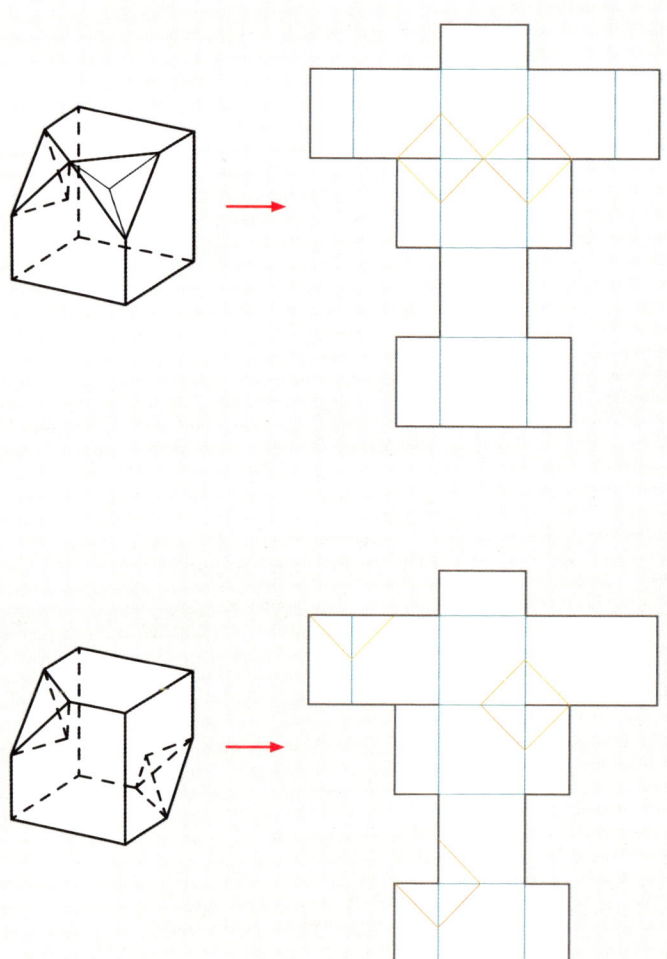

3.1.5 Pillow Curve

3.1.5.1

The Pillow Curve is so-called because it resembles the shape of a pillow. It can be used on any package as a substitute for a straight edge, replacing a straight fold with eye-shaped double curves. This simple replacement means it is remarkably adaptable, giving a softer, more decorative appearance to an otherwise straight-edged geometric form.

However, it is vital to understand how the exact radius of the curve is calculated and constructed, because a curve that is too flat or too bulbous will either be invisible to the eye or make a 3-D form too curved to lock shut.

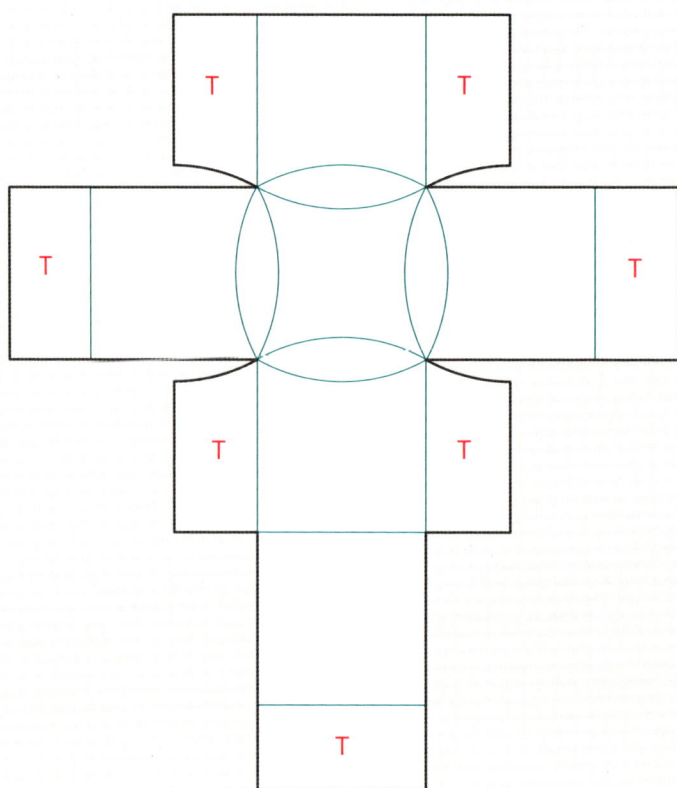

3.1.5.2

Note how the straight edges along the top of the cube are replaced with double curves. This can be done along any fold line. It can also be done around the perimeter of the net, though this can sometimes make a 3-D form difficult to lock neatly.

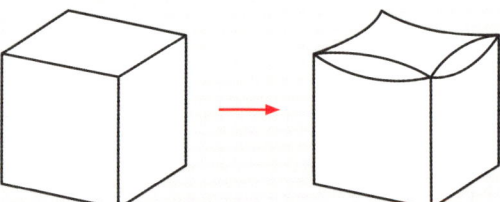

3.1.5.3

To construct a Pillow Curve, first draw a net for a conventional cube. Also draw center lines through every edge that will have a Pillow Curve, including those which will have curved tabs.

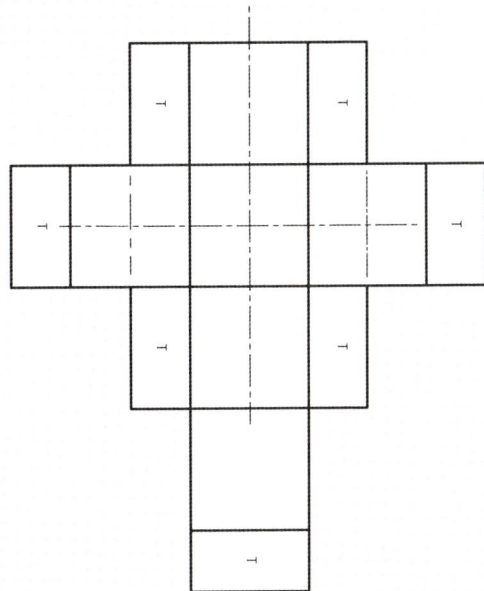

3.1.5.4

The straight edge marked in yellow will be substituted with a Pillow Curve. Measure its length and add 5%. Thus, for example, a 10cm edge, plus 5%, will be 10.5cm. This is the radius of the curve, marked in blue.

Stretch a pair of compasses to a radius of 10.5cm and place the needle on the central construction line that passes through the midpoint of the chosen edge, such that the pencil exactly touches one end of the line. Then, draw an arc from one end of the line to the other, here shown in red.

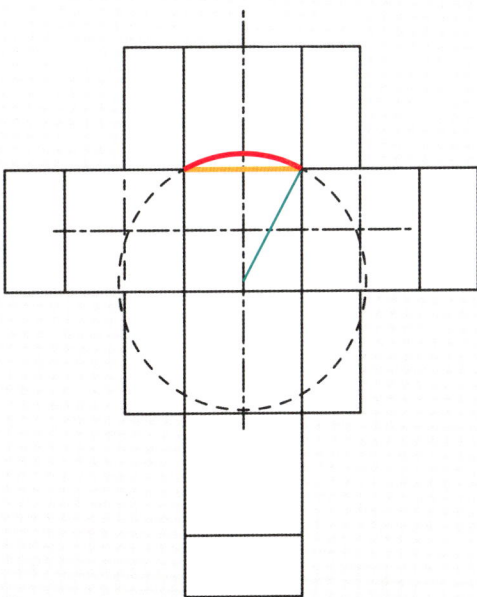

3.1.5.5

Repeat the previous step, but this time on the other side of the chosen line. Thus, the needle will be placed on the same construction line, but this time below the line. Draw another arc.

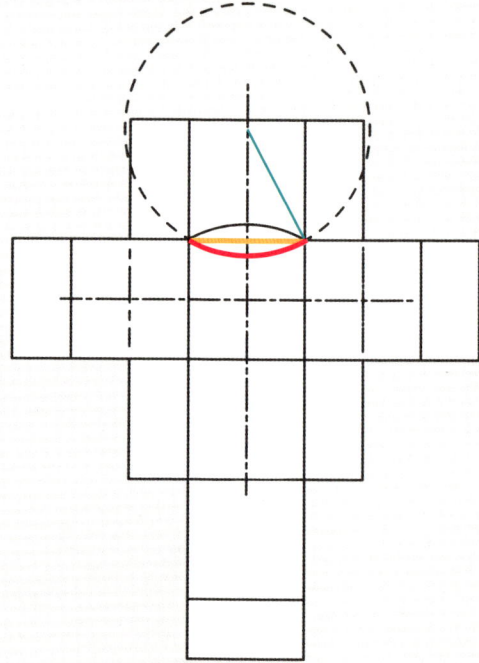

3.1.5.6

This shows the completed Pillow Curve along 1 edge.

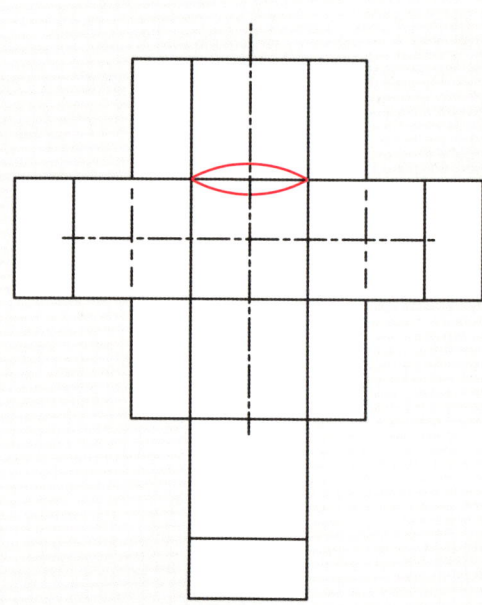

3.1.5.7

Repeat the previous steps as many times as necessary to complete the Pillow Curves. Remember to also draw small arcs at the edges of each affected tab.

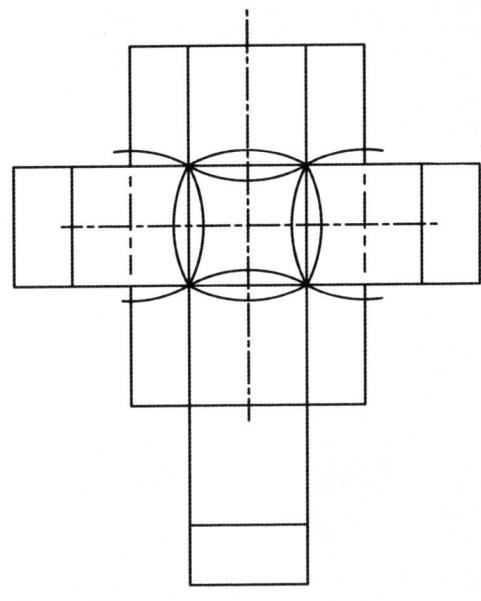

3.1.5.8

Remove all the construction lines. Importantly, the 4 straight edges substituted by the Pillow Curves must also be removed. They must not be creased.

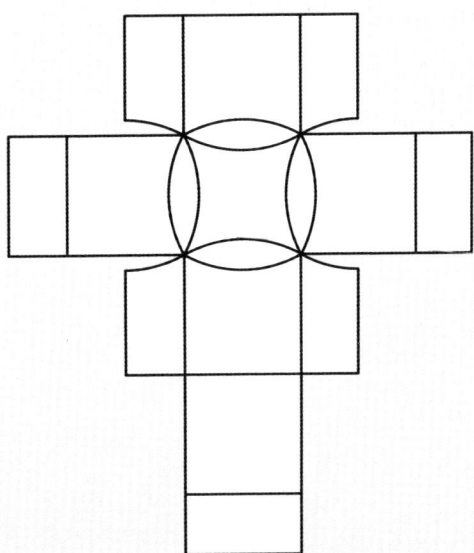

3.1.5.9

To make the curves by hand, first carefully cut out a piece of card shaped like a wedge of pie. The curves must be longer than any on your design. Cut the curved edge with extreme care, so that it is smooth and regular. If you can do this with a mechanical aid, that is preferable to cutting by hand.

3.1.5.10

Lay the curved card exactly over the path of one of the curves and make the crease with the back of a cutting knife, as though making a conventional straight crease. Repeat for all the other curved creases.

Note that the card must only be creased by the knife between the red arrows.

Cut and fold the remaining creases and edges in the conventional manner.

When folding the card into 3-D form, make the curved folds very carefully, just a centimeter at a time. They cannot be made roughly.

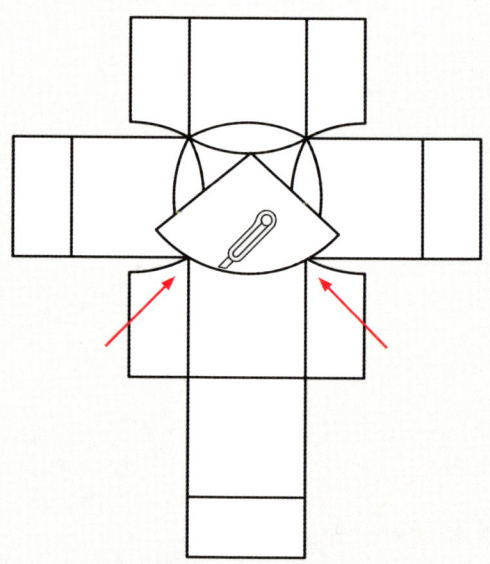

3.1.5.11

Any straight edge can easily be substituted with a Pillow Curve, in any combination. Pillow Curves can also be created shorter than a full edge, or be doubled on an edge, even trebled.

To construct these unusual curves, measure the start and end points of each curve, then add 5% of the length to find the radius of the curve.

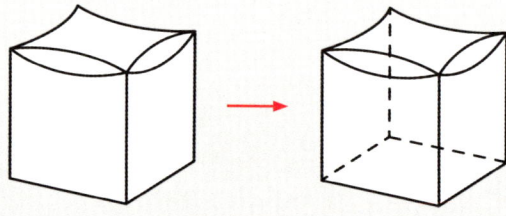

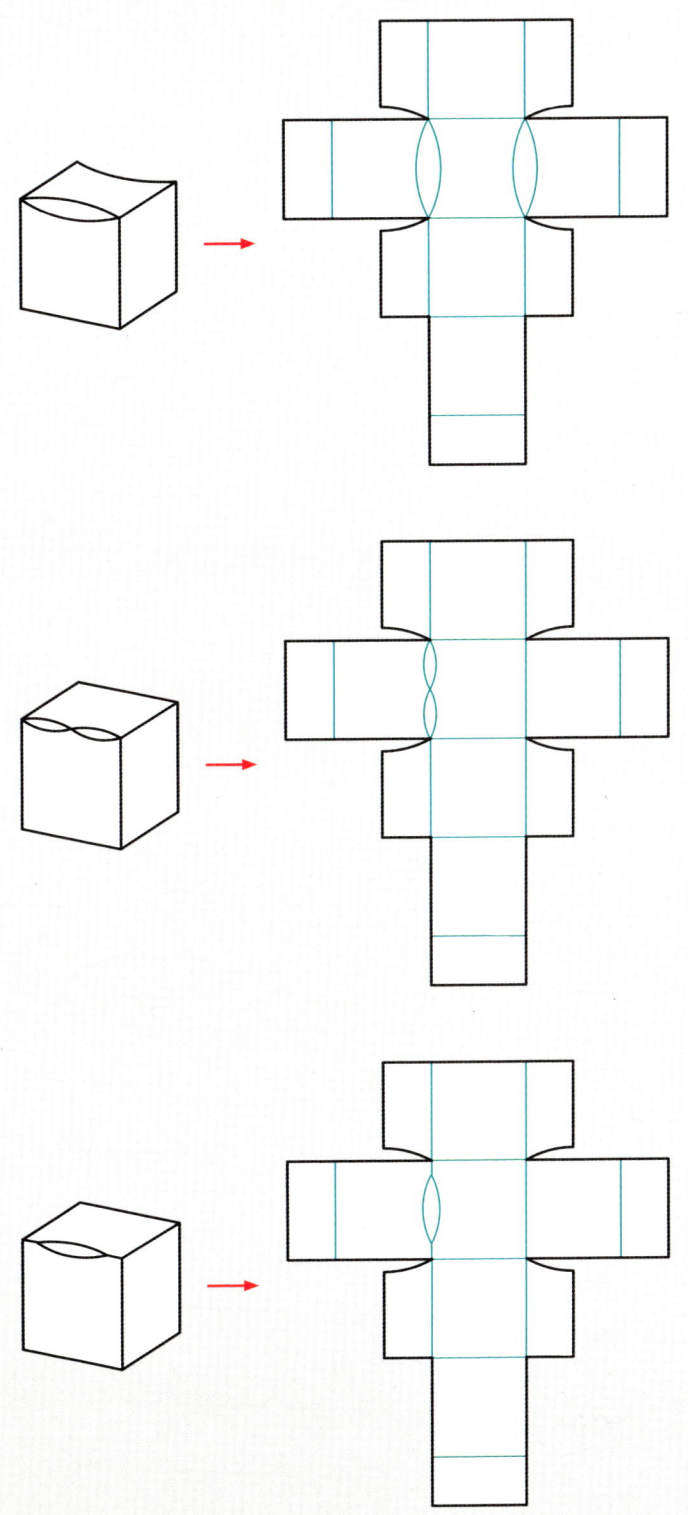

3.1.6 Skewed Cube

3.1.6.1

This is one of the more basic, less showy cube distortions, but is nevertheless – and perhaps because of its plainness – a very practical form. The angles of the rhombus are the choice of the designer. When constructing it, be aware that angle "a" plus angle "b" equals 180-degrees.

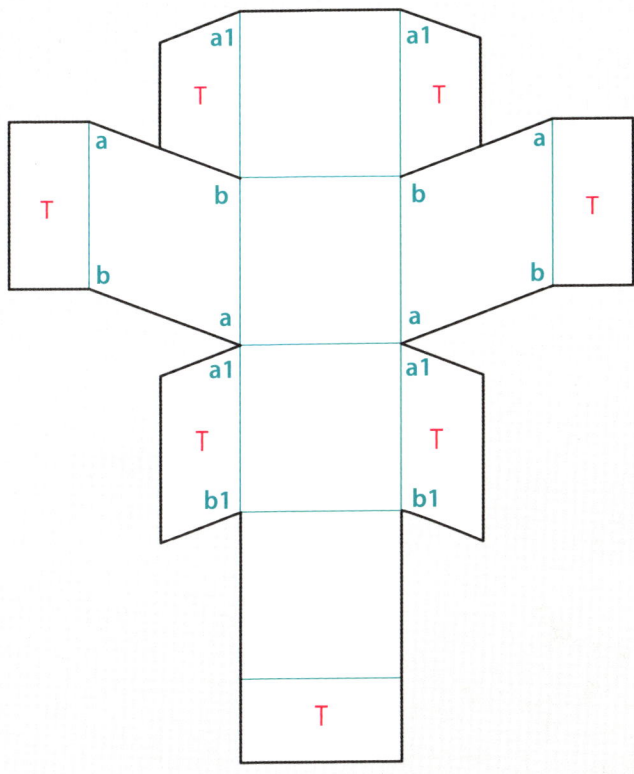

3.1.6.2

The Skewed Cube is simply a conventional cube being pushed a little to one side, so that 2 of its faces become rhombi.

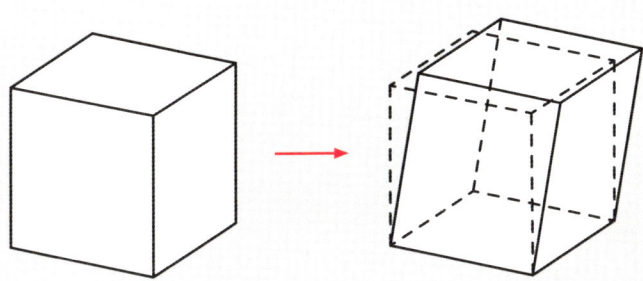

3.1.7 Rhomboid Box

3.1.7.1

Although seemingly a complex and crazy net, comparison with the small cube net shows that its structure is very conventional. This particular configuration was chosen in preference to a more familiar cube net because it creates extra space around the perimeter for the tabs.

Note that all the faces are rhombi, and all the angles are either 60- or 120-degrees. There are no right-angled corners!

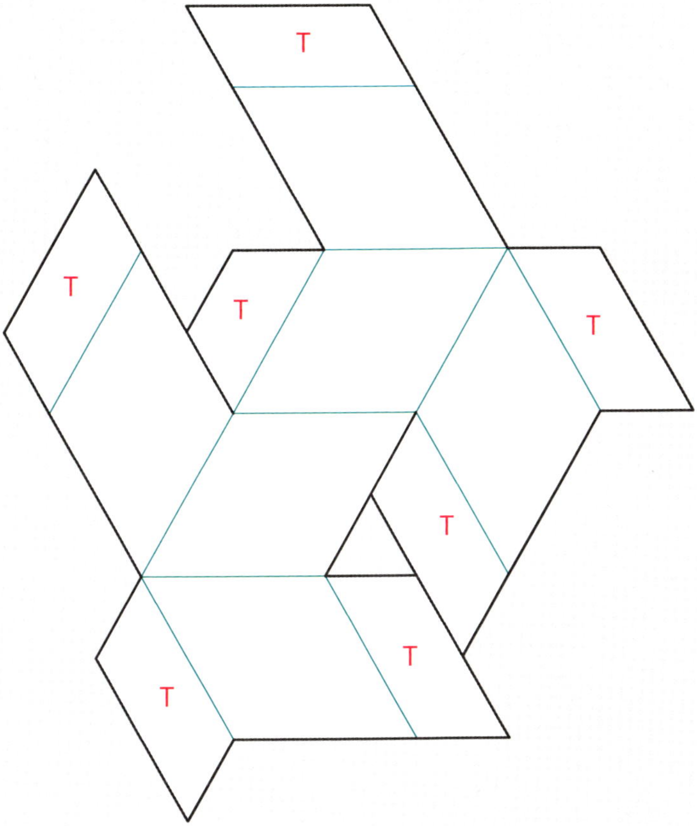

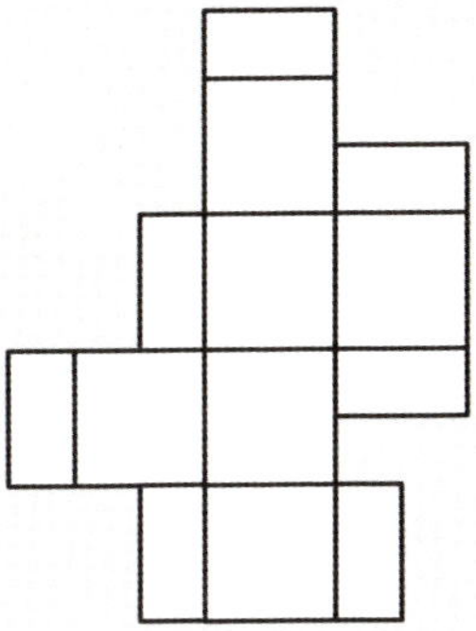

3.1.7.2

Imagine a line connecting opposite corners of a cube, then stretching that line. The 6 squares of the cube would distort to become rhombi.

The resulting form looks like a cube in distorted perspective. It's a rather disturbing effect, very difficult for the brain to assimilate and also difficult to photograph.

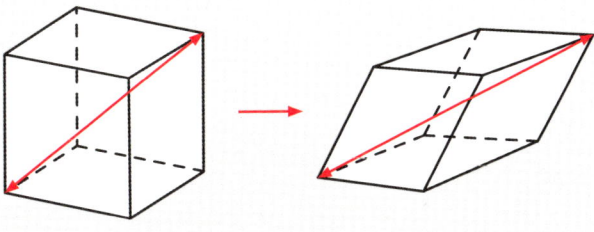

3.1.8 Faceted Box

3.1.8.1

The triangles are all isosceles. The sides of equal length could be the same length as the squares, but they could also be longer or shorter, thus making the resulting form taller or shorter than a cube.

Note the flanges which will help hook the small tabs into the folded box and lock it tight shut.

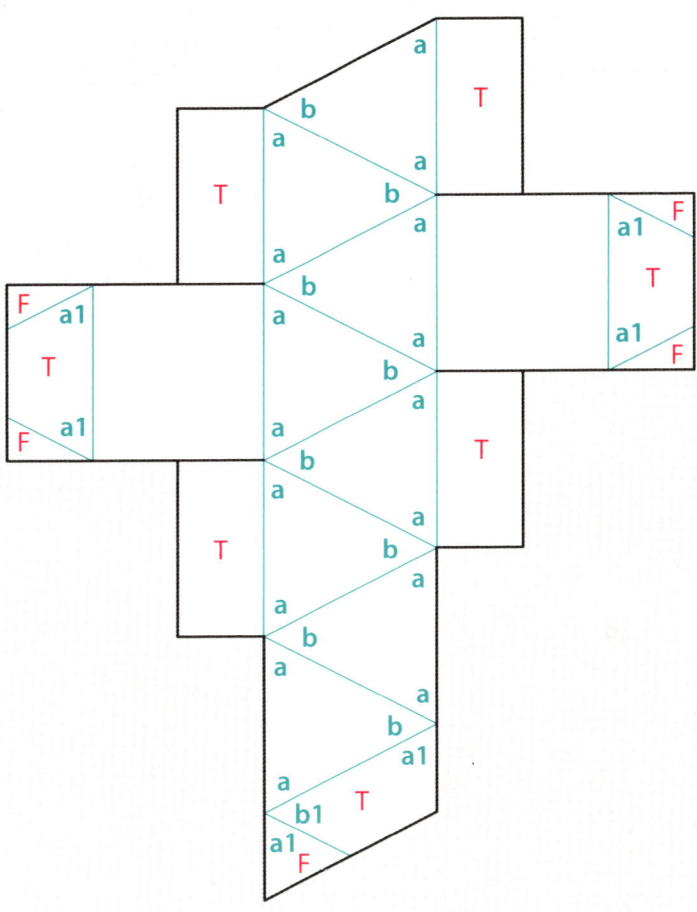

3.1.8.2

Look at the 2 square faces. The top square is rotated 45-degrees compared to the bottom one. When this is done, the 4 vertical edges divide to become 8, creating a series of triangles around the sides.

The result is a very beautiful form, which in any light will have many different tones across the triangles.

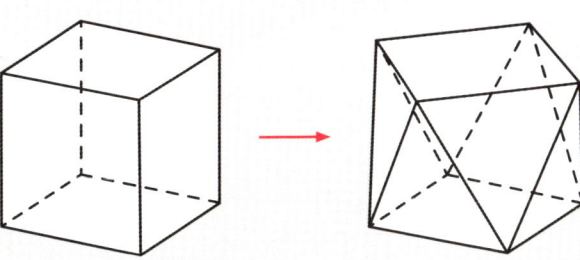

3.1.9 Pinched Box

3.1.9.1

Note the repetition of angles "a" and "b" and the 2 wedge-shaped tabs. Measure angle "b" to determine the "b1".

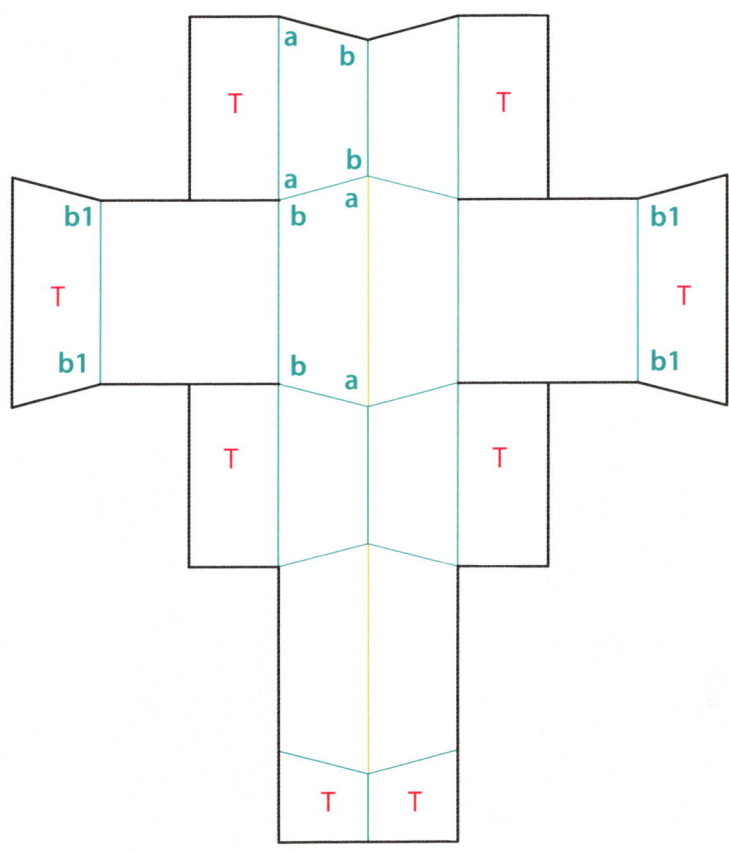

3.1.9.2

Unusually, the Pinched Box distortion features 2 valley folds. The box has the same surface area as a cube, but if a cube was pinched (squeezed) in the area of the valleys, it would distort to become both concave and convex, as shown.

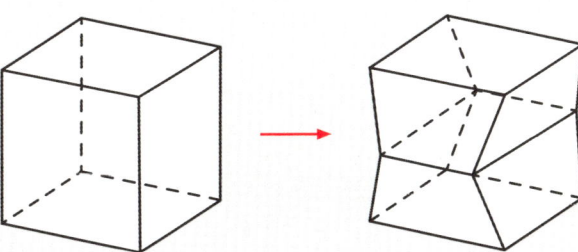

3.1.10 Twisted Box

3.1.10.1

4 rhombi are stacked down the middle of the net, flanked by 2 squares. Angle "a" plus angle "b" totals 180-degrees. For the twist to form successfully, angle "a" must be around 100-degrees and thus, angle "b" will be around 80-degrees. If the angles is too large or too small, the twist will not work.

Note how the tabs on the left and right edges of the net are not square.

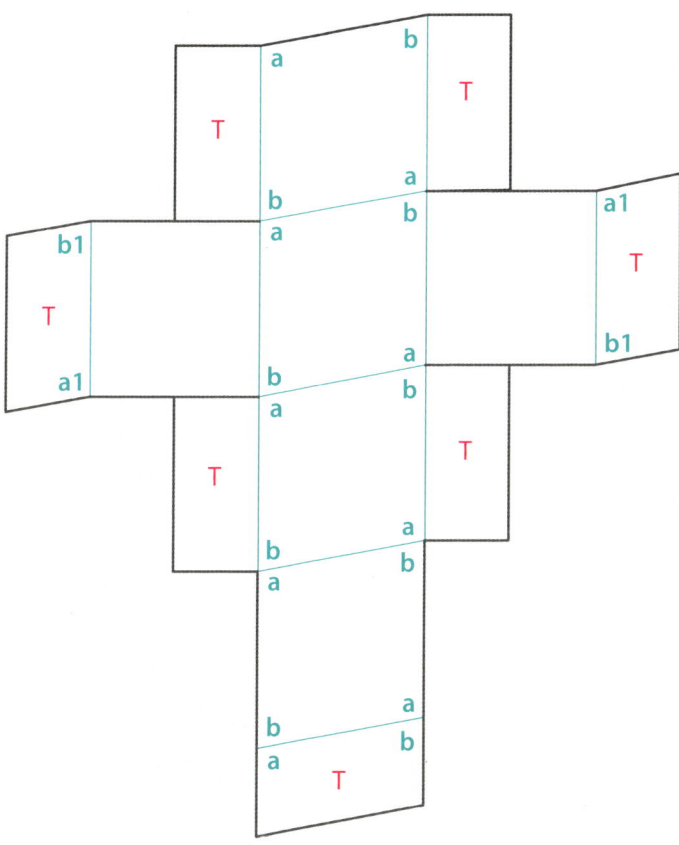

3.1.10.2

Of all the ways to distort a cube, this is perhaps the most beautiful. The curved faces create subtle shadows, and the subtle twist makes the 2 square faces offset in relation to each other.

However, the twist will not work if the box is taller, or if the angles of 100- and 80-degrees become – for example – 70- and 110-degrees.

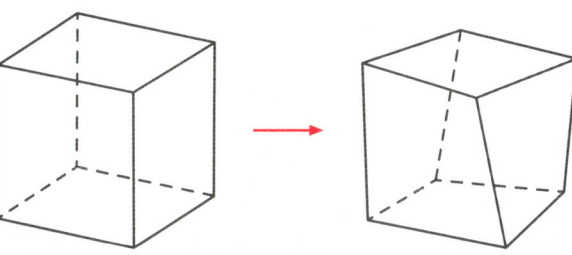

3.1.11 C-curve Box

3.1.11.1

The curves down the center of the net are made in the same way as the curves of the Pillow Curve (see 3.1.5) and likewise have a radius of the straight-line distance between the ends of the curve, plus 5%. Please refer to the Pillow Curve description to see how to create, draw and make the C-curves.

Note how adjacent curves are arcing in opposite directions. Thus, the box is a variation of the Pinched Box (see 3.1.9), substituting curves for straight edges and flat polygons.

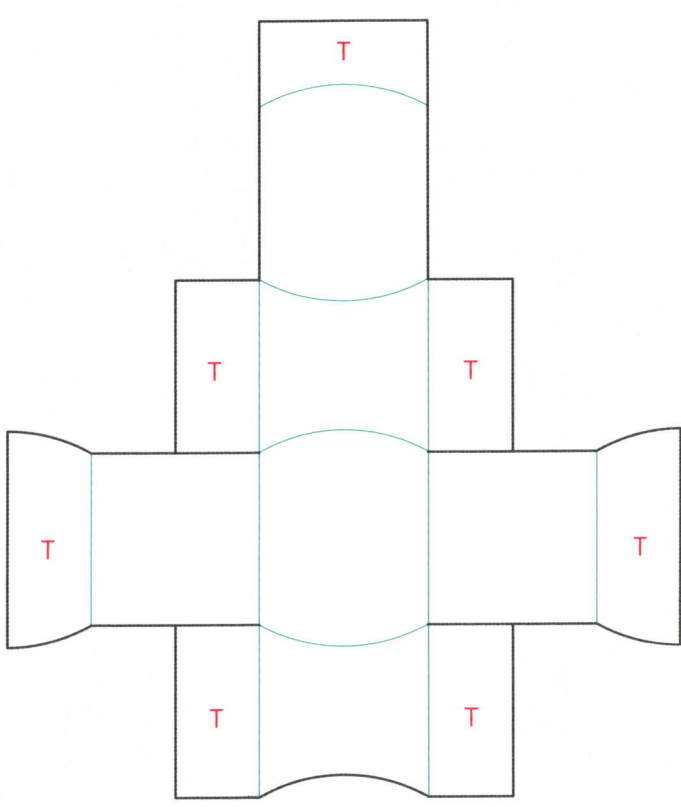

3.1.11.2

The concave and convex surfaces create subtle effects of light and shade. The mirroring of the curves around the sides means that this technique can be used not only on boxes with 4 sides, but also on hexagonal and octagonal boxes. However, as the number of sides increases, the curves become less and less dramatic. The mirrored curves cannot be made on boxes with an odd number of sides, such as 5, 7 or 9.

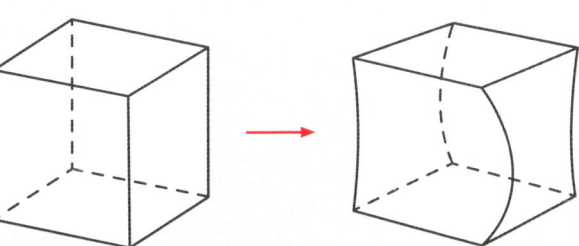

3.1.12 Truncated Pyramid Box

3.1.12.1

The 4 trapeziums around the sides are all equal. Angle "a" plus angle "b" equals 180-degrees. Calculate these angles, then create the same angles for tabs "a1" and "b1".

When folding up the box, everyone will try to lock in the tab on the small square, last of all, as though it's a lid. However, being wedge-shaped, the tab will not lock into the box. Instead, the sequence for closing the box should see that tab locked in early, then the tabs at the left and right of the net locked in at the end. These tabs are 90-degrees and will lock in easily.

The 4 large tabs are each half of a trapezium.

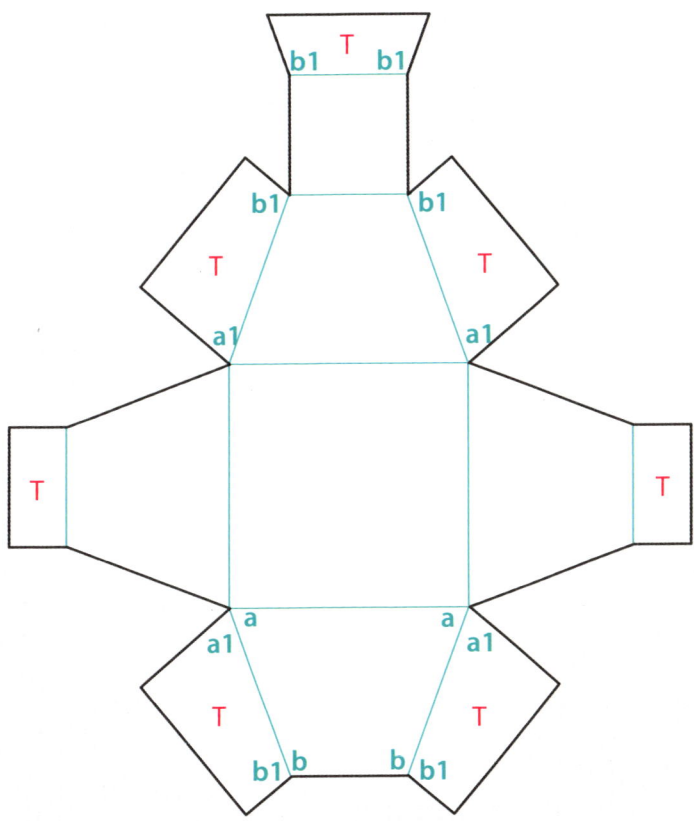

3.1.12.2

The Truncated Pyramid Box is one of the most versatile and attractive ways to distort a cube. Many variations can be made, stretching and squashing the trapeziums, or creating a box with 3, 5, 6 or more sides.

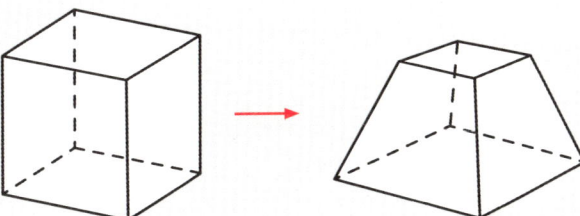

3.2 Strategies for Creating Multi-packs

We have seen in 3.1 how to distort a cube by stretching, twisting and shaving it, but another way is to create a multi-pack by dividing a cube into equal units. These units will pack together to create a cube (or another simple polyhedron of your choice), but are also packages in their own right.

The method to create these units is fascinating. It can also become esoteric, sculptural and ultimately, impractical, but even the simple dividing of a cube can create beautiful and surprising forms that every designer should consider.

Multi-packs offer the opportunity to create unique forms that pack together without any waste of space.

3.2.1 Faces, Edges and Corners

All flat-plane polyhedrons have faces, edges and corners. Other polyhedrons, such as cylinders or cones, have curved faces and edges, but sometimes no corners.

3.2.1.1

Here are the faces, edges and corners, identified and named. In geometry, a corner is more correctly called a "vertex" ("vertices" in the plural), but because this is a book for beginners, the more common word "corner" is used, which is also correct.

- A face is a 2-dimensional plane.
- An edge is an 1-dimensional length.
- A corner is a 0-dimensional point.

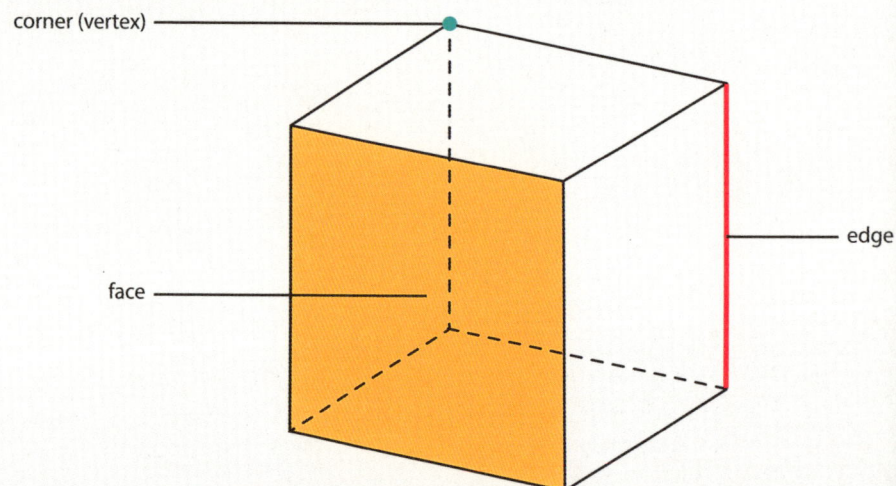

3.2.1.2

Here are the faces on a cube. There are 6 of them, all the same shape and size. They are all squares.

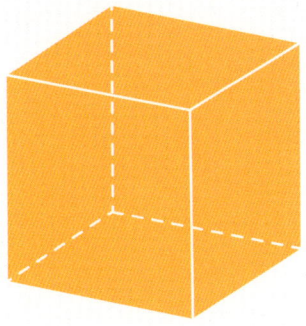

3.2.1.3

Here are the edges on a cube. There are 12 of them. They are all of equal length.

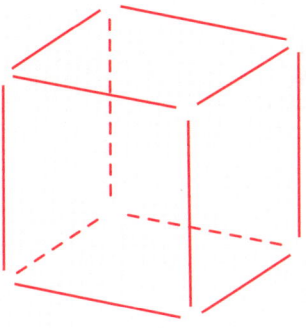

3.2.1.4

Here are the corners on a cube. There are 8 of them, equally spaced along the edges and across diagonals.

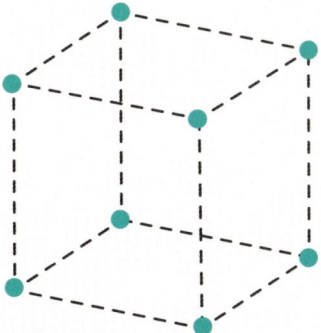

When dissecting a cube, it is common to use either faces, or edges or corners as the basis for the dissection. It is rare to mix these 3 different elements into 1 dissection system.

So, the next section will examine in turn, how each of these elements can be used. It is a fascinating and beautiful journey!

3.2.2 Dissection Using the Faces System

3.2.2.1

Here is one of the faces.

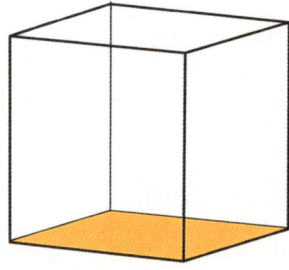

3.2.2.2

Since there are 6 faces, we want to create a polyhedron that will occupy one-sixth of the volume of the cube and thus, one-sixth of the surface (a square face). The only point in or on the cube common to all 6 polyhedrons, will be the center-point of the cube. Thus, all 6 polyhedrons must meet at this point.

When the center-point is connected to the corners of a face, the shape generated is a 4-sided pyramid, with its apex at the center of the cube.

6 of these pyramids will create a cube.

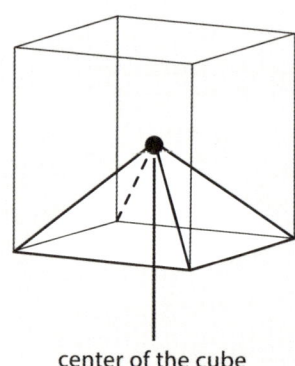

center of the cube

3.2.2.3

This is the pyramid.

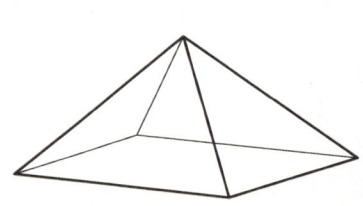

3.2.2.4

Using the Pythagoras Theorem and eventually trigonometry, the exact shape of the triangular face of the pyramid can be calculated.

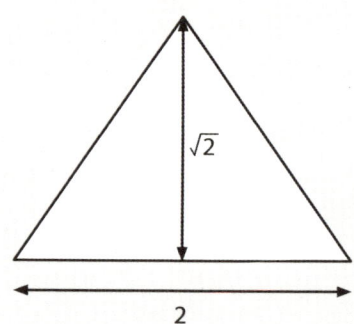

Unit: cm

3.2.2.5

This is the net for the pyramid. Note how, to create extra strength, all the triangles are connected at the apex.

The tabs are added using the system described in the previous chapter. The Flanges ensure that the tapering tabs have extra width and will hook themselves into the net.

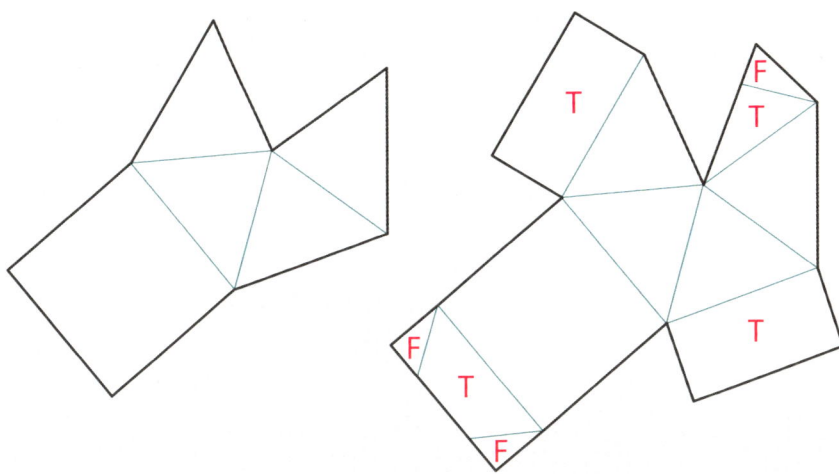

3.2.2.6

This is where the face system of dissection becomes creative.

A simple pyramid which is one-sixth of the volume of a cube, can be cut symmetrically in half, so that each resulting unit is one-twelfth of the volume.

The pyramid can be cut in half along a diagonal, or from the middle of 1 edge to the middle of the opposite edge. These forms will be different, but all are one-twelfth of the volume of the cube.

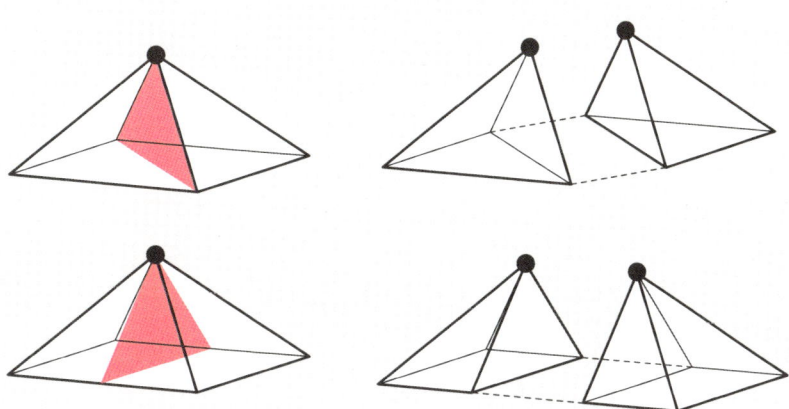

QUESTIONS

1. Make 2 pyramid units. How many ways can you glue them together to create a meta-unit that is one-third of the volume of the cube? Create a one-piece net for it. What other ways will these units tessellate in 3-D?

2. Similarly, make 3 pyramid units. How many ways can you glue them together to create a hyper-unit that is half the volume of the cube? What other ways will these meta-units tessellate in 3-D?

3. You have made sub-units, each one-twelfth of the volume of the cube. How many different ways can you glue 2 of them together to create a unit one-sixth of the volume of the cube, but not a pyramid?

4. Can you glue 3, 4 or 6 of these sub-units together to create meta-units that are one-quarter, one-third or one-half of the volume of the cube?

5. Many meta-units can be cut in half and reassembled in a different way. Try to do this.

3.2.3 Dissection Using the Edges System

3.2.3.1

Here is 1 of the edges.

An edge is 1 dimensional and does not over any of the surface of the cube. Since there are 12 edges, we need to use the edge to generate a shape that will cover one-twelfth of the surface.

A little experimentation will show that this surface-shape is half on one face and half on the other. Both halves are 45-, 45- and 90-degree triangles. They are connected by an edge, which is a line of symmetry across the surface-shape.

12 of these surface-shapes will cover the surface of the cube.

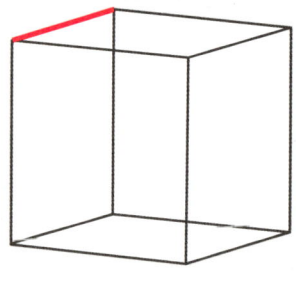

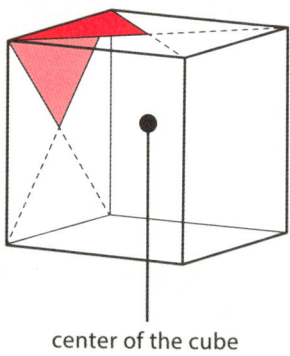

center of the cube

3.2.3.2

Since there are 12 edges, we want to create a polyhedron that will occupy one-twelfth of the volume of the cube. The only point in or on the cube common to all 12 polyhedrons will be the center-point of the cube. Thus, all 12 polyhedrons must meet at this point.

When the center-point is connected to the corners of the surface-shape generated in the previous step, the form created is pyramid-like, with 6 triangles.

12 of these polyhedrons will create a cube.

3.2.3.3

Here is the form.

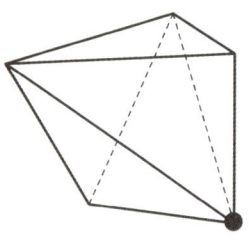

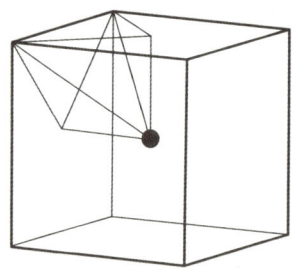

3.2.3.4

Using the Pythagoras Theorem and/or trigonometry, the exact shape of the triangular faces can be calculated.

Note those 6 right-angled corners.

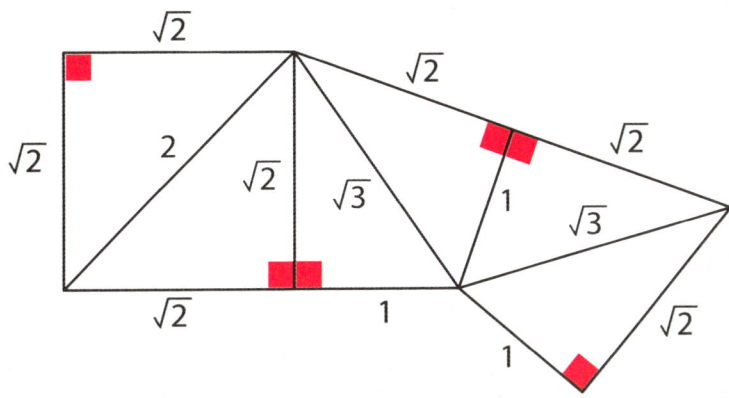

3.2.3.5

This is the net for the form.

The tabs are added using the system described in the previous chapter. The Flanges ensure that the tapering tabs have extra width and will hook themselves into the net.

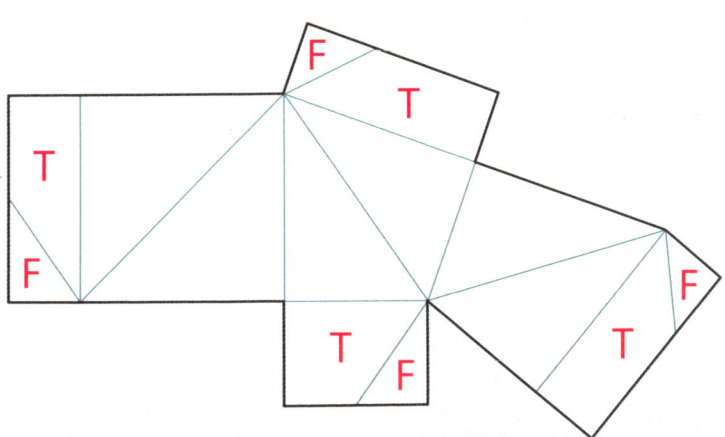

The pyramid-like form, which is one-twelfth of the volume of the cube, can be combined with others to create meta-units.

QUESTIONS

1. Make 2 pyramid-like units. How many ways can you glue them together to create a meta-unit that is one-sixth of the volume of the cube? Create a one-piece net for it. What other ways will these units tessellate in 3-D?

2. Similarly, make 3 pyramid-like units. How many ways can you glue them together to create a hyper-unit that is one-quarter of the volume of the cube? What other ways will these meta-units tessellate in 3-D?

3. Repeat, but glue 4 or 6 pyramid-like units together. Will all these meta-unit assemblages combine to create a cube, or can other forms also be made? It is astonishing how many options can be found!

4. How many ways can these meta-units be cut in half and reassembled in a different way? Try to do this.

3.2.4 Dissection Using the Corners System

3.2.4.1

Here is one of the corners.

A corner has 0 dimension and does not over any surface of the cube. Since there are 8 corners, we need to use the corner to generate a shape that will cover one-eighth of the surface.

A little experimentation will show that this surface-shape is 3 squares, each of which covers one-quarter of a square face. The squares all meet at the corner.

8 of these surface-shapes will cover the surface of the cube.

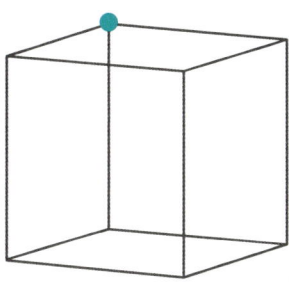

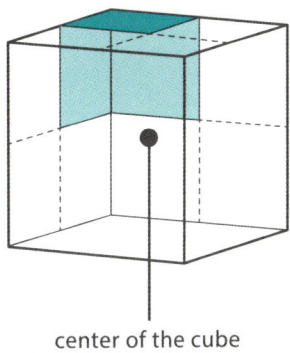

center of the cube

3.2.4.2

Since there are 8 corners, we want to create a polyhedron that will occupy one-eighth of the volume of the cube. The only point in or on the cube common to all 8 polyhedrons, will be the center-point of the cube. Thus, all 8 polyhedrons must meet at this point.

When the center-point is connected to the corners of the surface-shape generated in the previous step, the form created is a simple cube. It's almost a disappointment!

8 of these small cubes will create a cube.

3.2.4.3

Here is the form. The dot identifies the corner at the center-point of the cube.

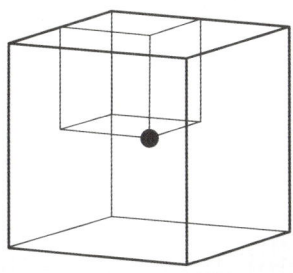

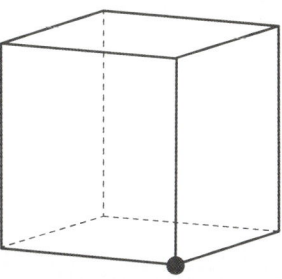

3.2.4.4

It is possible to divide a small cube into 2 symmetrical halves, so that each sub-unit becomes one-sixteenth of the volume of the cube. There are many interesting ways to do this. Here is one way, dividing the cube into 2 triangular prisms.

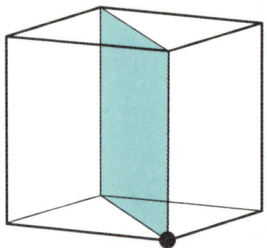
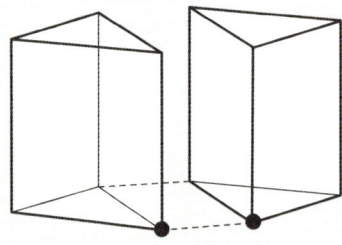

3.2.4.5

Because the polyhedron we are dissecting is a small cube (which has one-eighth of the volume of the large cube), we can likewise dissect the large cube in the same way, to create 2 equal halves.

This creates a fractal system of dissection, that can create multi-packs directly from the large cube, or by dividing the cubic corner unit.

Here are two further ways to dissect the cube – either the large cube, or the small corner unit.

These units can themselves be further divided and then recombined in many ways.

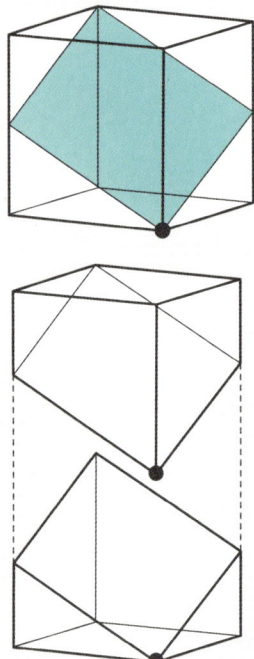
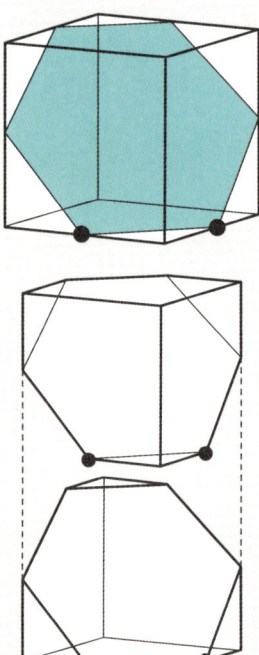

Triangualar prism

QUESTIONS

1. In how many ways can you divide a cube into 2 equal volumes? Is there a rule or a system?

2. What happens when these halves are recombined to create forms that are not cubic, but which have the same volume as a cube?

3. How many times can you dissect a cube into equal halves, then recombine the halves in a different way, then dissect again, then recombine again...and so on? Is there a limit?

3.2.5 Summary

6 faces, 12 edges and 8 corners immediately give the designer 3 different systems for dissecting a cube into differently-shaped multi-packs, based on 3 different numbers (6,12 and 8). These numbers can themselves be further divided to create 12, 24 and 16 multi-packs, which can be recombined in different ways.

Alternatively, the original 6, 12 and 8 units can be combined together to create meta-units with double, treble, quadruple or sextuple the volume of the original unit...which can then be dissected and recombined in different ways.

From this, it can be understood that designing multi-packs is a lifetime's study!

Of course, there are many practical considerations that will limit the playful exploration of three-dimensional geometry, but nevertheless, this is a vast subject beyond the scope of this introductory book to explore in depth. It is a topic where package design meets three-dimensional geometry, meets sculpture.

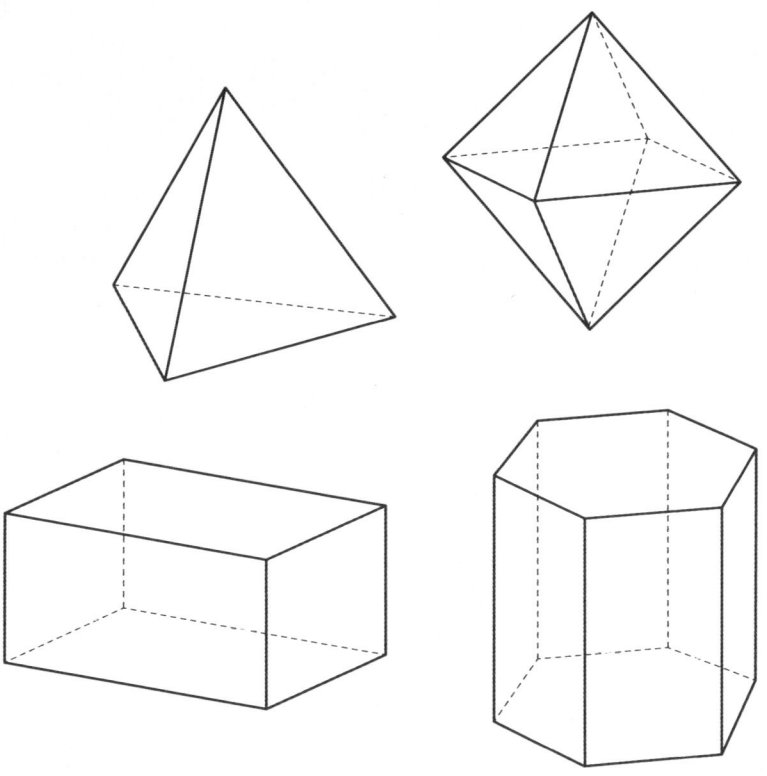

The cube is without doubt the easiest polyhedron to divide into multi-packs. However, other polyhedrons should not be dismissed. Each of the examples shown here has a unique combination of faces, edges and corners, providing 3 different methods to divide them into many different and exciting multi-pack forms.

Cubes are very practical, but they can also be a tyranny!

04
LOCKS

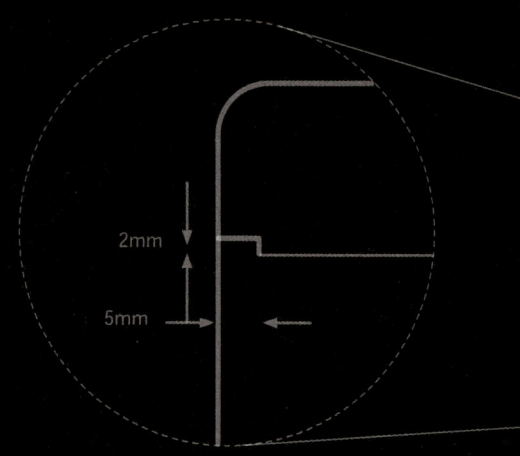

P156 **4.1 Click Lock**

P159 **4.2 Tongue Lock**

P161 **4.3 Crash Lock**

P165 **4.4 Hexagonal Crash Lock**

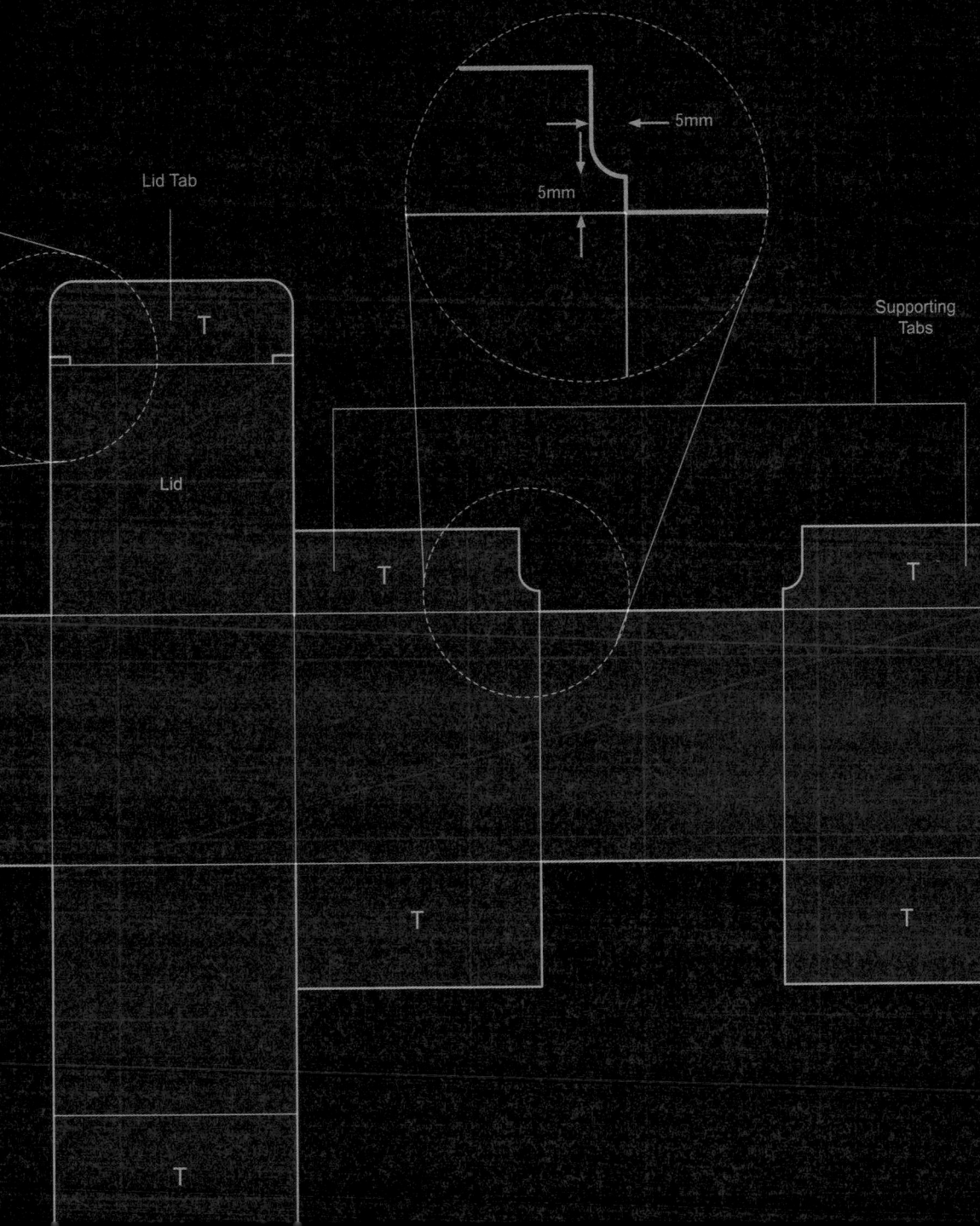

There are infinite number of ways to lock any volumetric solid. The only limitation is the imagination of the designer. If you want to design on your own, try to create different ways to lock a face, an edge, or a corner, with 1 lid-part, 2 lid-parts, or more, symmetrically or asymmetrically, with or without a glue line... so on. It is a vast and esoteric subject, without an end!

The function of a lock is to secure a package. If the package is one-use only, then the lock may be torn open. However, if the package is to be opened and closed many times, then it needs to be designed in a wholly different way so that the closure will always remain strong, despite repeated use.

There are two basic ways to create locking lids that can be repeatedly opened and closed. They are the "Click Lock" and the "Tongue Lock", both explained in this chapter. Which one is preferable depends on the weight of the material you are using and the weight of the contents.

Additionally, the "Crash Lock" and "Hexagonal Crash Lock" are included here, to show how the base of a box can be significantly strengthened, so that it may contain a greater weight than a simple Click or Tongue could support without breaking and spilling the contents.

4.1 Click Lock

The Click Lock is the most common lock in packaging, and for good reason. It is easy to construct, and lock securely and can be opened and closed many times without damage.

It is best suited for use with thinner cartons that have some flexibility. Heavier, stiffer materials do not always respond well to a Click Lock.

Thinner cartons generally hold relatively lightweight contents, so the Lock is mostly used on smaller packages.

The drawing here focuses only on the design of Click Lock. The shape of the box and the way the base Lock must be designed separately.

4.1.1

The exact structure of the central lid tab and the 2 supporting tabs is crucial to the success of the Click Lock. Please construct it carefully, paying close attention to the measurements, curves, the placement of the folds and so on.

The size of the cube is unimportant, but this demonstration has a side length of 6-7cms.

Done well, the Click Lock will close with a satisfying "Click" (of course!) and will take some persuasion to open.

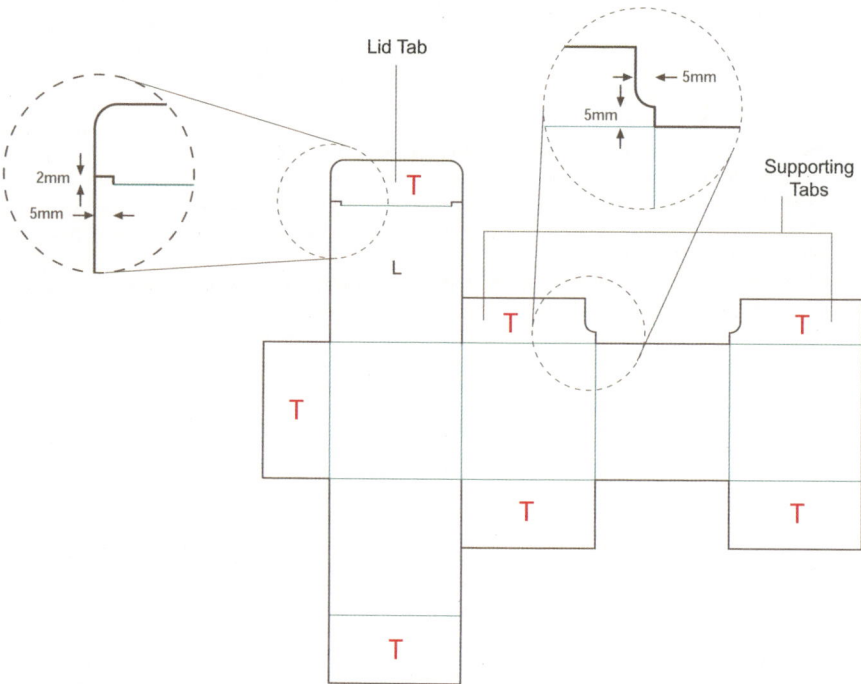

4.1.2

The Click Lock is perfect for closing lightweight cuboid packages. Less well-known is how good it is at closing problematic polyhedrons with tabs of less than 90-degrees. These tabs tend to fall out of a locked box, so a way needs to be found to hold them in.

It has previously been suggested to use Flanges, that is, pieces added to the sides of tabs to help hooking them into a box. They work well, but not for a lid that needs to be repeatedly opened and closed.

Here is a simple net for a tetrahedron. The 3 tabs are all the correct shape but will not lock the 3-D form securely shut. The addition of a Click Lock will solve the problem.

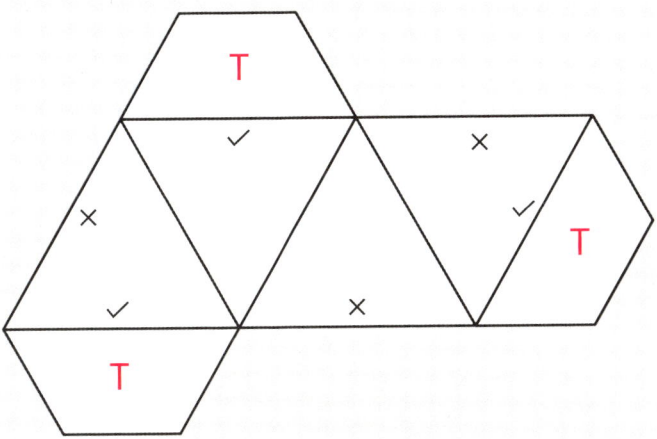

4.1.3

This is the net seen in 4.1.2 adapted for a Click Lock. The tab on the right will "click" into the other 2 tabs and hold itself tight shut.

The compromise to this solution is that the clean edges of the tetrahedron are interrupted by 2, small protruding tabs. On the one hand, they show where the form should be opened; on the other hand, they perhaps look a little unsightly.

What is beyond doubt is the practicality of the solution.

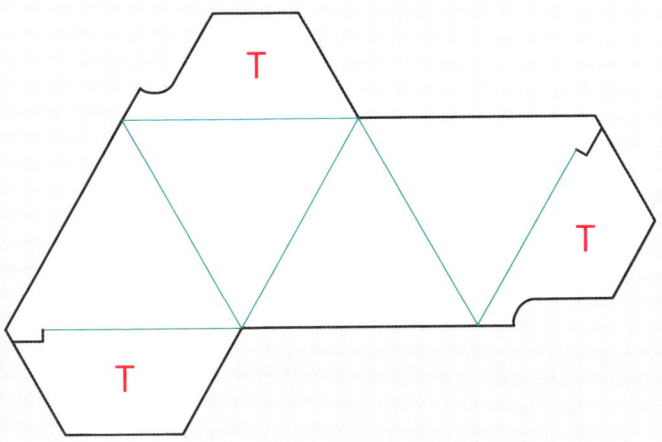

Pyramid with Click Lock

4.2 Tongue Lock

The Tongue Lock is stronger than a Click Lock and can be used on heavier grades of carton or plastic. It will lock an edge very securely shut in a neat and unobtrusive way.

4.2.1

As before, it must be made with great attention to detail. The width and length of the tongue are flexible, but the slit into which it is inserted must be exactly the same as the width, to ensure a tight fit.

For extra strength, the tab on the top side of the net can be glued.

In the drawing, GT means Glue Tab.

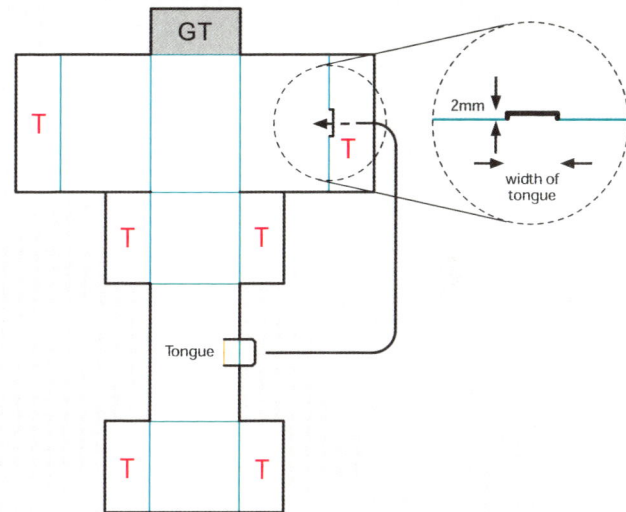

4.2.2

Tongue Locks can also be used to lock polyhedrons with triangular faces. Note that in this case, the Locks must be added to every edge – there is no designated lid face.

Any edge on any polyhedron, can be locked in this way.

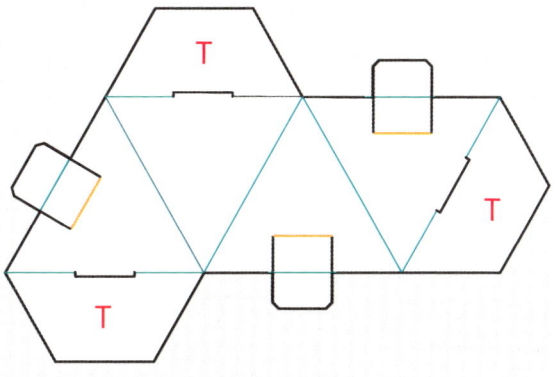

Pyramid with Tongue Lock

4.3 Crash Lock

A Crash Lock will substantially strengthen the base of any package with a square or rectangular base.

4.3.1

Compare the top of this net with the base. The top shows a conventional configuration of tabs, whereas the base shows a Crash Lock. If the net was folded and a product placed inside, it would quickly fall through the conventional end, but would be easily supported by the Crash Lock.

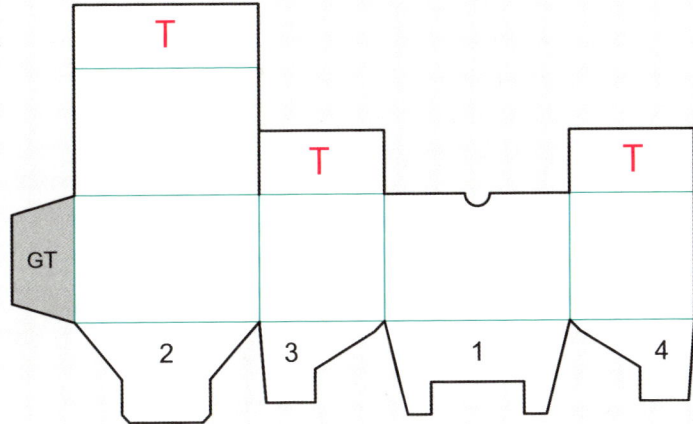

4.3.2

To design a Crash Lock, begin by measuring the dimensions of the base and dividing it as shown. The position of the yellow dots is crucial to the design of each of the 4 pieces.

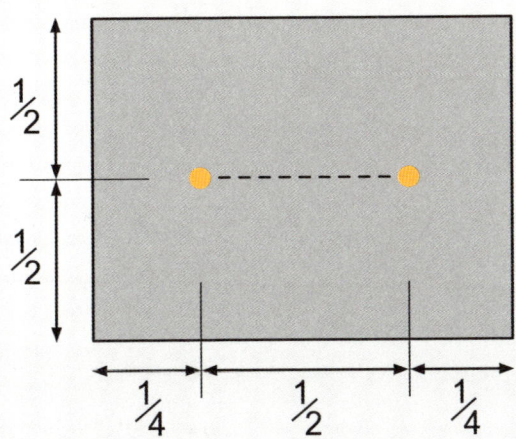

4.3.3

Piece 1.

Design it as shown, paying particular attention to the position of the yellow dots.

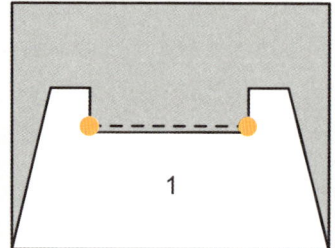

4.3.4

Piece 2.

As always, pay close attention to the position of the yellow dots.

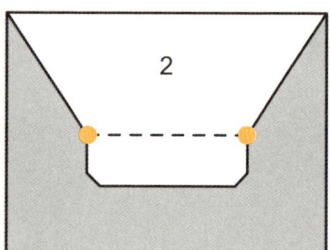

4.3.5

Pieces 3 and 4.

They are the same shape, but mirror images of each other.

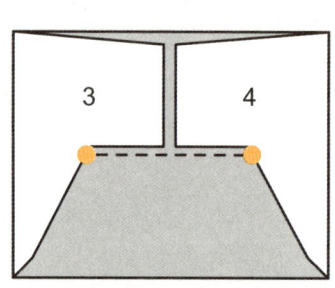

4.3.6

This shows the distribution of the 4 pieces around the bottom of the box.

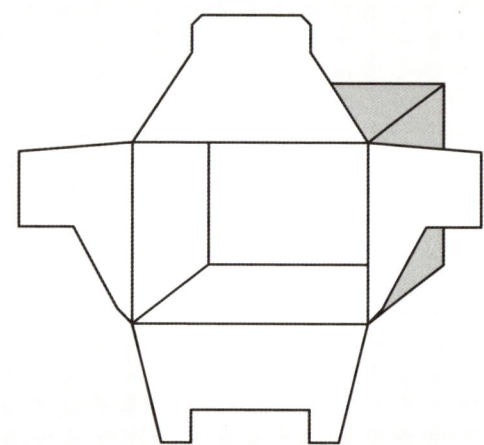

4.3.7

To lock the base, first fold in the bottom piece...

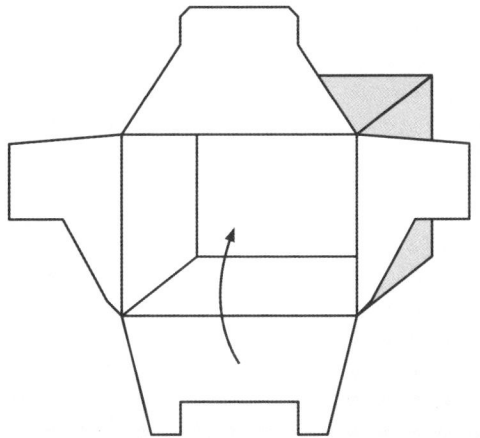

4.3.8

Then the left and right pieces...

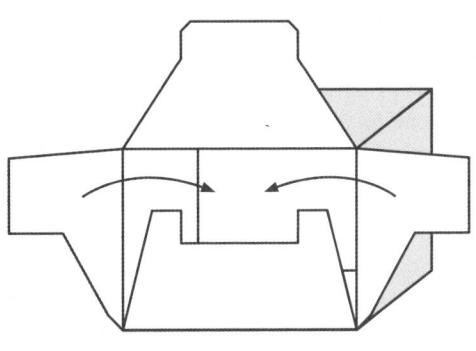

4.3.9

...and finally, slide in the top piece.

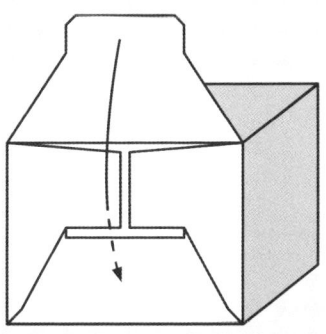

4.3.10

Ideally, the base will lock tightly, with no separation of the pieces in the positions of the yellow dots (the dots are not shown). However, if made too perfectly, the final base piece will have no space to slide inside. If this is happening, shave off slivers of card on pieces 1, 3 and 4, to create a narrow slot across the middle. This is one of those instances in which perfect measurements need to compromise a little!

Made well, the base will support a considerable weight.

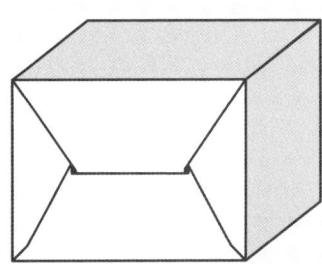

Cuboid with Crash Lock

4.4 Hexagonal Crash Lock

A useful variation of the Crash Lock for a rectangular base, is a Crash Lock for a hexagonal base.

The pieces need to be designed with the same care as before.

4.4.1

This is the complete net. Note the contrast in structures between the hexagon and tabs along the top edge and the Crash Lock pieces along the bottom edge.

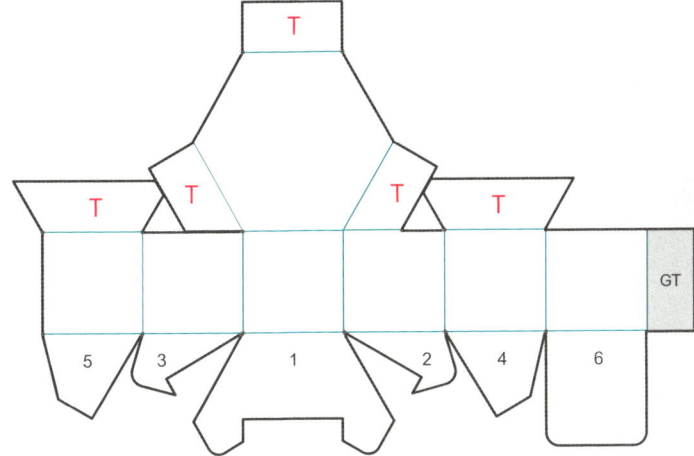

4.4.2

First, locate the position of the 2 yellow dots. When this has been done, the shape of the 6 different pieces can be designed, each in relation to the dots.

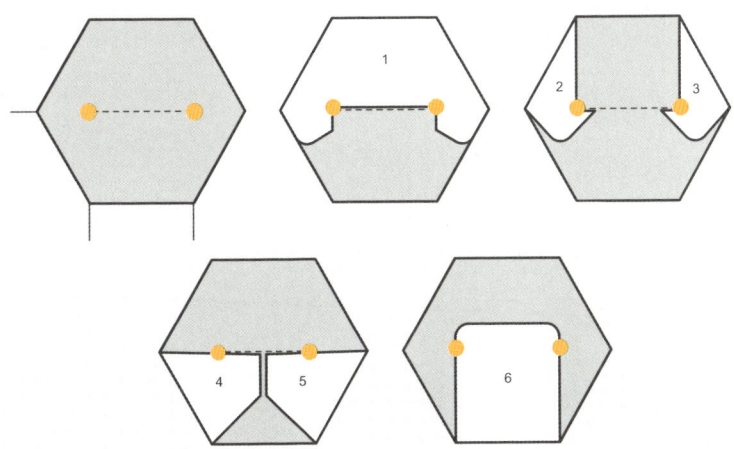

4.4.3

Fold in the 6 pieces in the order shown. The final piece may not slide in if the other pieces have been cut very accurately, so shave off slivers from some or all of them to create a narrow slot across the middle.

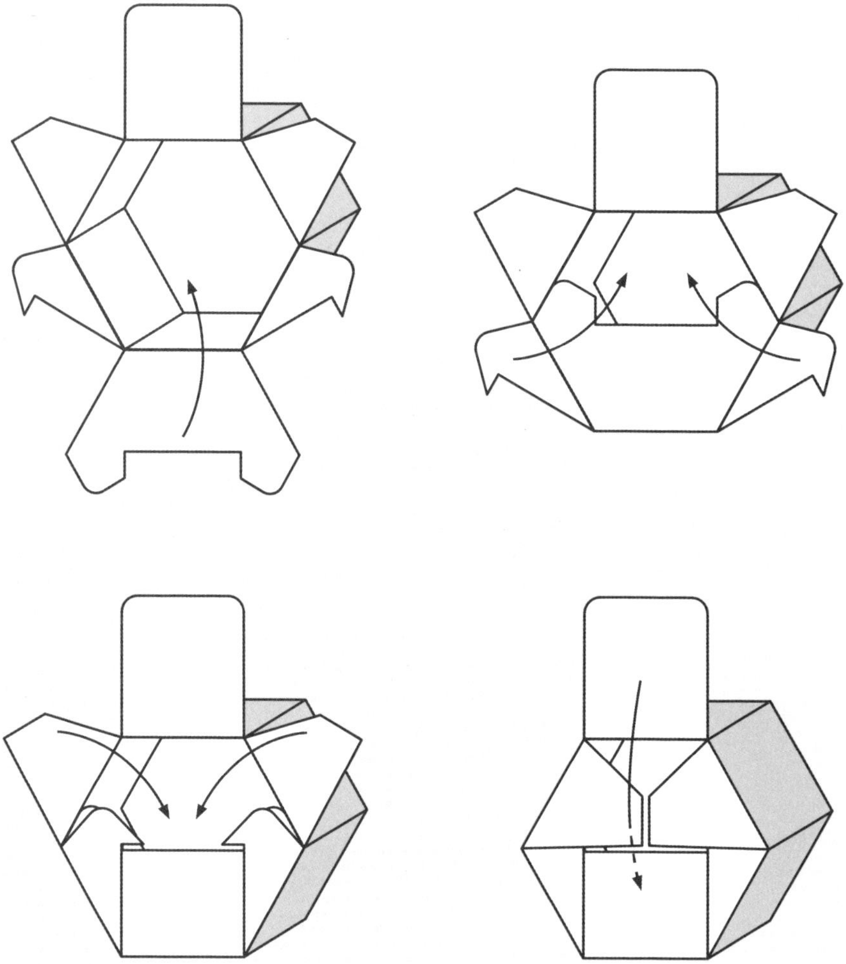

4.4.4

This is how the Hexagonal Crash Lock will look. It's a very elegant structure, satisfying to make. It will also hold a huge amount of weight!

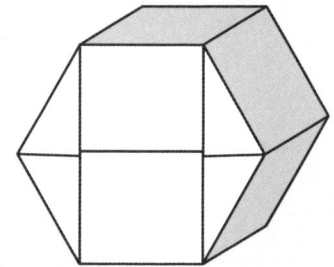

Hexagonal Crash Lock

CREATIVE PACKAGING...
AND MORE

P172 **5.1 Sample Nets**
P186 **5.2 Student Gallery**
P194 **5.3 Beyond Packaging**

If you have carefully read the previous chapters and – it is hoped – made as many examples as possible, you will have understood how to make a strong, self-locking volumetric solid and will have learnt strategies for designing your own.

For 30 years, I have run projects for students of Design and for professionals, in which they created their own package designs, having learnt the basics in a day-long workshop. This chapter presents some of their works.

In today's screen-based studios, making packaging by hand can be a novel and sometimes stressful experience for some students, unused to measure by hand, using a sharp knife and generally needing to work with a high degree of manual precision. However, after some understandable hesitation, they would quickly settle, and most would eventually create something beautiful and perfectly designed.

With packaging, you know when something is right. The box will hold together tightly and is, in a sense, "perfect". With many other aspects of design, perfection is an illusion. It is a matter of opinion if a typeface is right, if a color is appropriate and if many other design choices are the best of all options, but with packaging there is only perfection and then progressive degrees of weakness. This strong sense of "right and wrong" in package design creates tight guidelines for an inexperienced designer, providing a clear path and giving a strong sense of personal satisfaction when a "perfect" design is achieved.

The work that follows has been created by students from Schools of Design across Europe and the USA. Each package design is the product of an intensive 4–5-day workshop and project, run under my supervision.

Many of these designs are NOT intended to be mass-produced commercial packaging. They are examples of the students exploring the net design system I taught, understanding 3-D geometry better, understanding how to lock volumetric solids, understanding how to lock a lid...and more, so that in a practical packaging project, designing something strong, functional and practical, but with a creative twist, becomes relatively simple and fun. Very importantly, taking time to understand the net design system deeply, means they can focus on the "Why" of their design choices, not the "How". The "How" is easy, the "Why" is infinitely more complex!

5.1 Sample Nets

These nets were created by students from Aalto University in Helsinki, Finland, made as part of an interdisciplinary Pack Age module. Some of the students were Design students, but others were from Marketing, Food Technology and other courses. The designs presented here are typical of their project work.

After 5 days working with me to create these nets, they would begin a series of real-world package design briefs, using their experience with me to guide them in their design works.

5.1.1

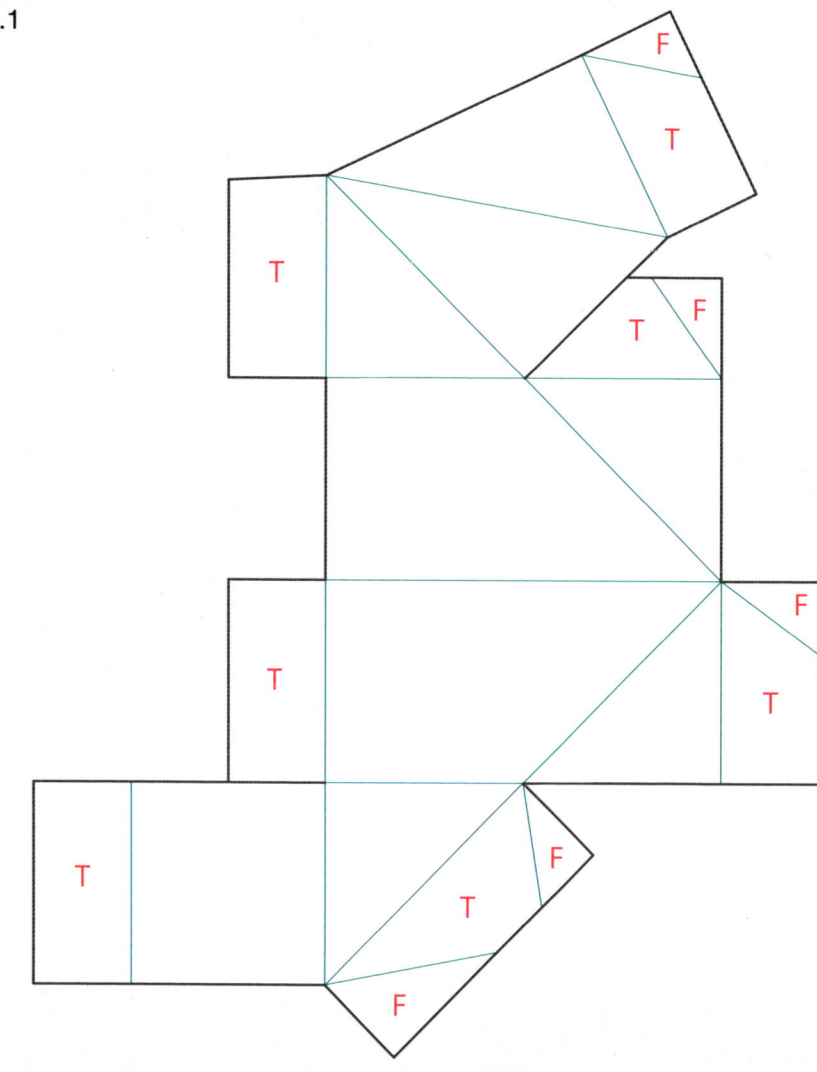

5.1.2

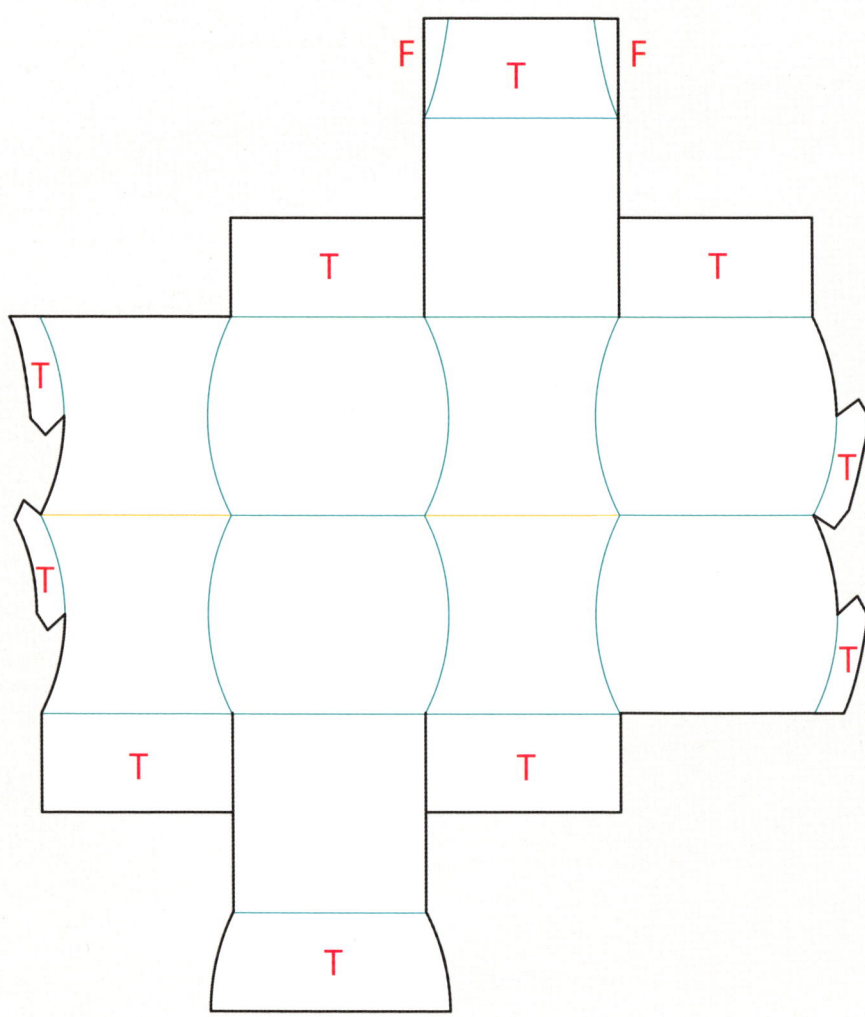

5.1.3

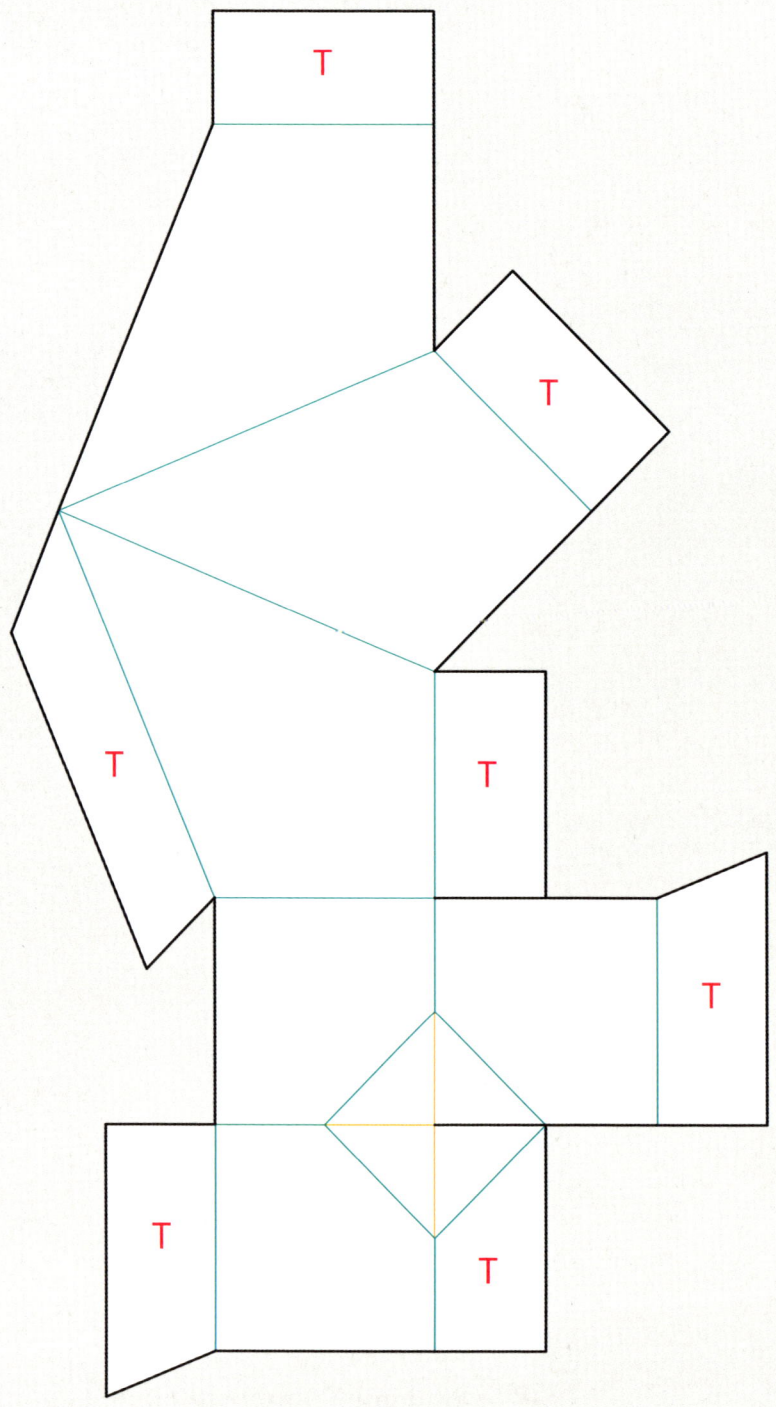

5.1.4

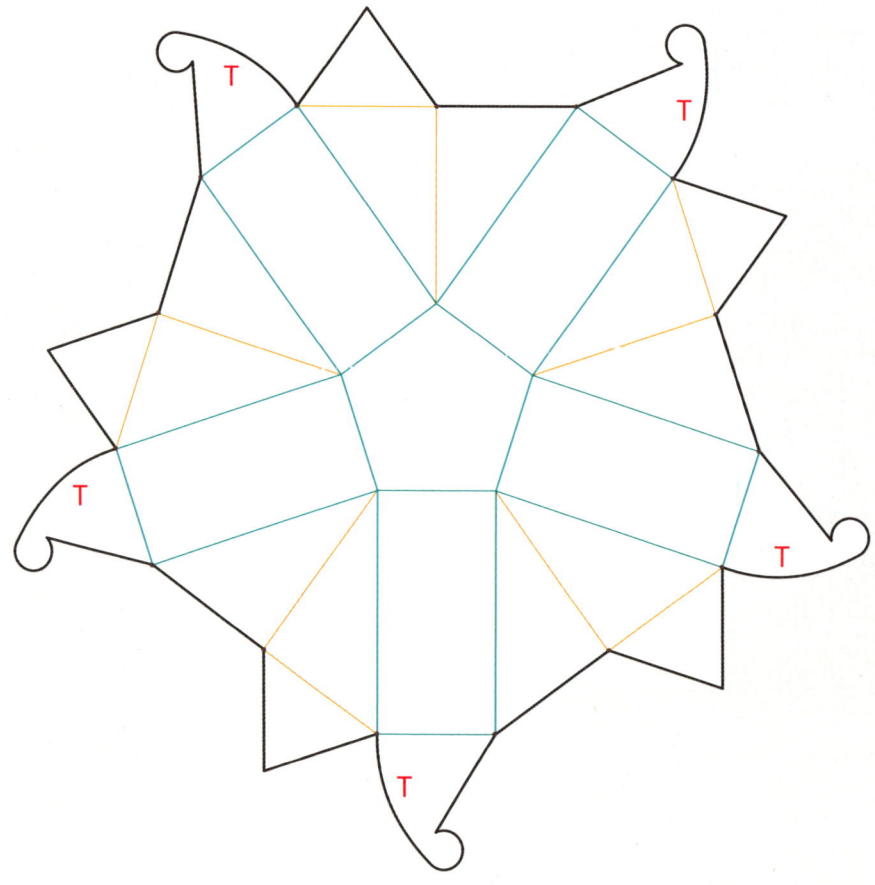

5.1.5

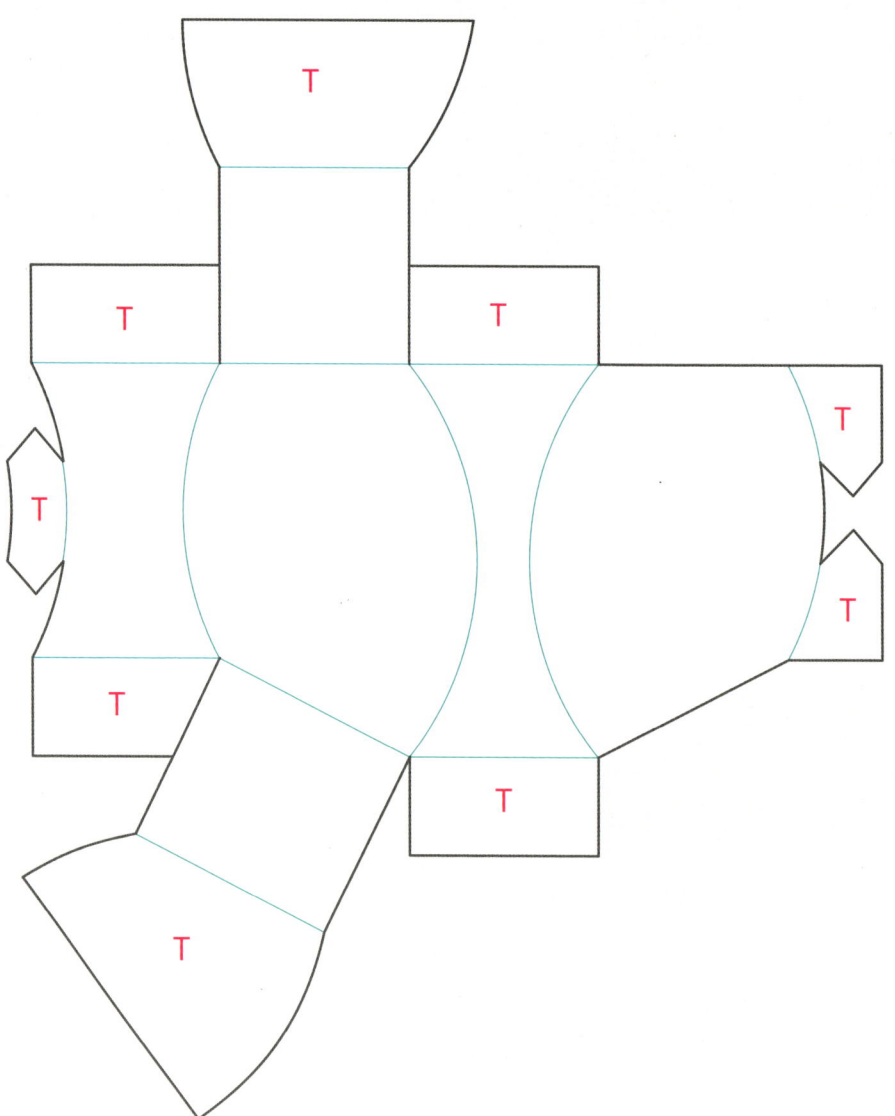

5.1.6

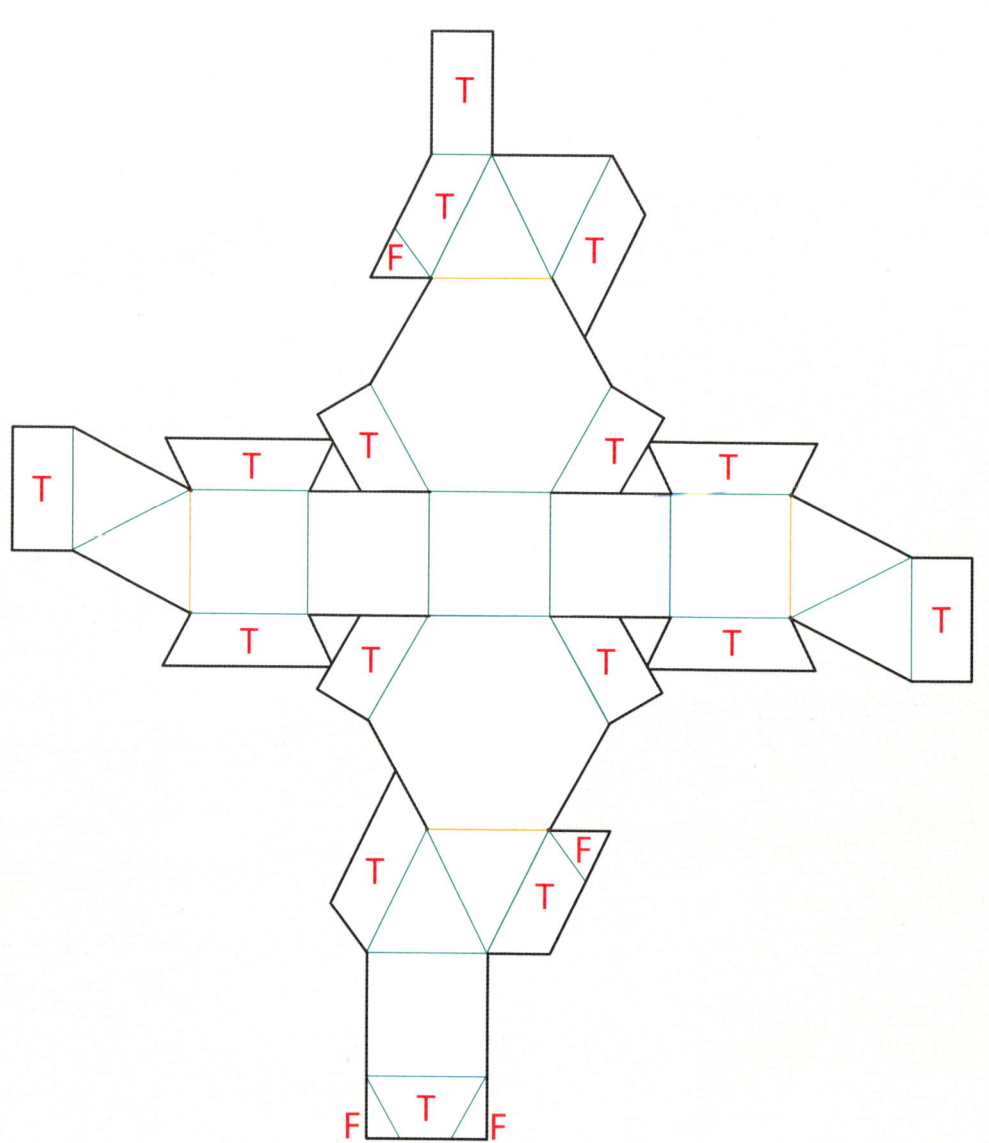

5.1.7

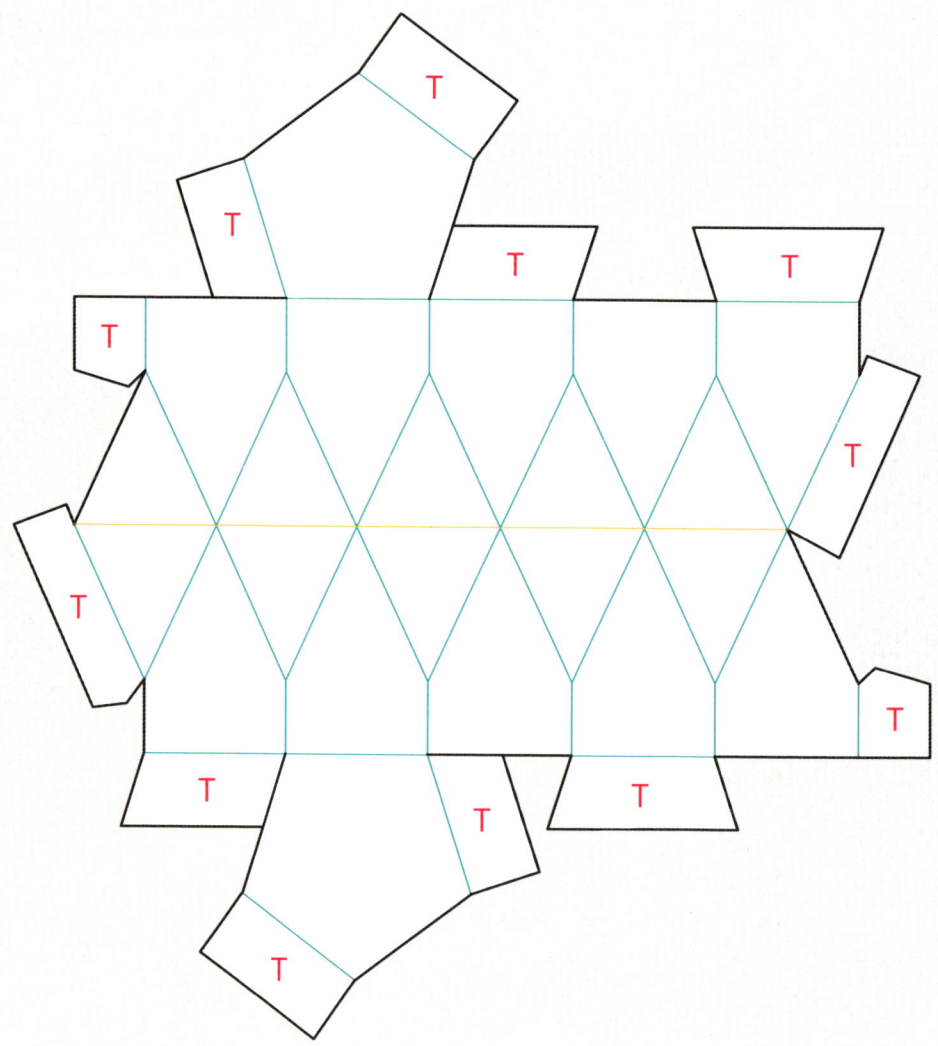

5.2 Student Gallery

The photographs on these pages were taken during my projects, often by myself using an iPhone. A few were taken later, by studio technicians. Most of the photographs were taken next to a large window, allowing the soft natural light to illuminate the different facets.

Almost all the packing is made by hand from 1 sheet of card, without glue.

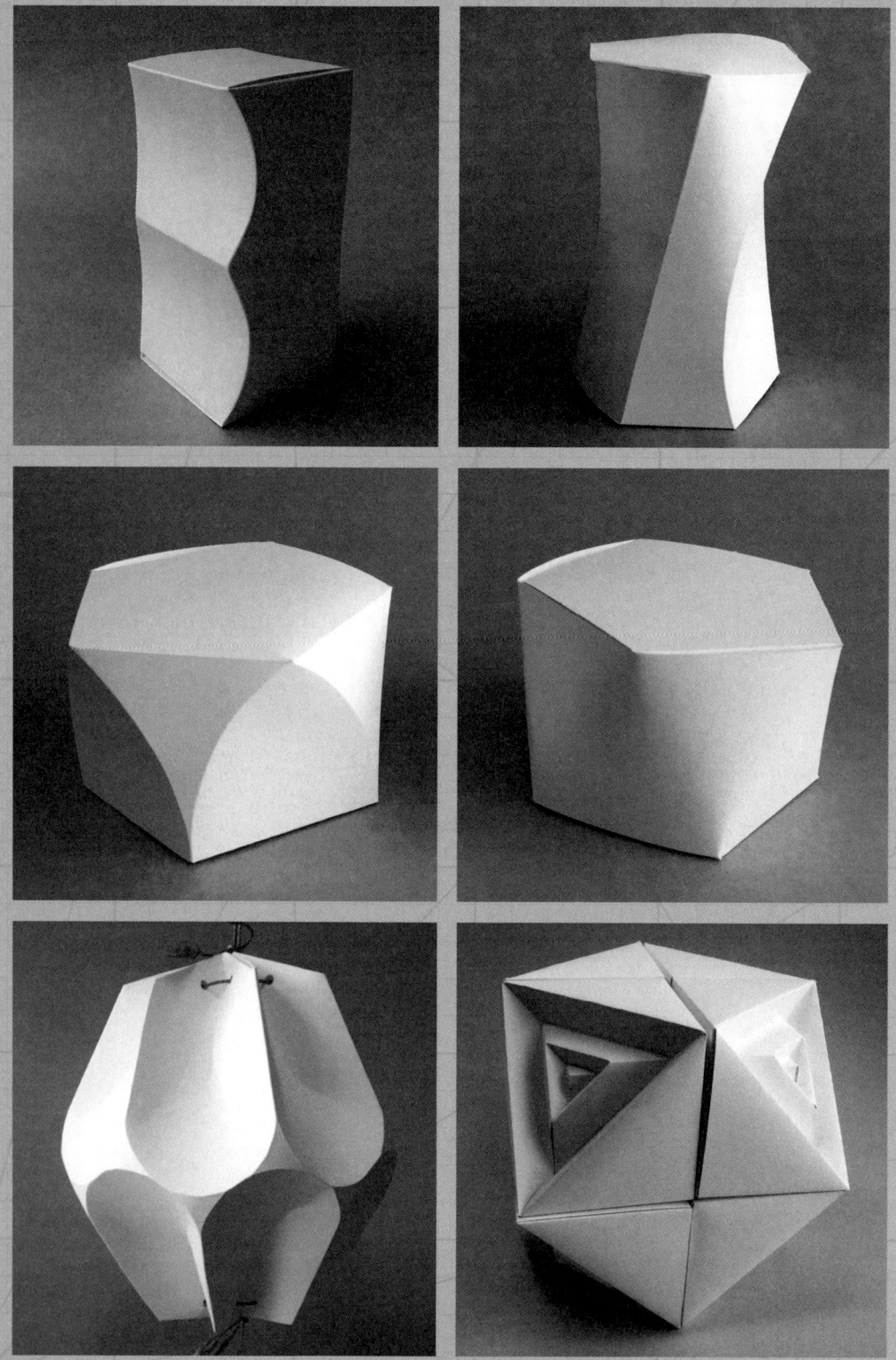

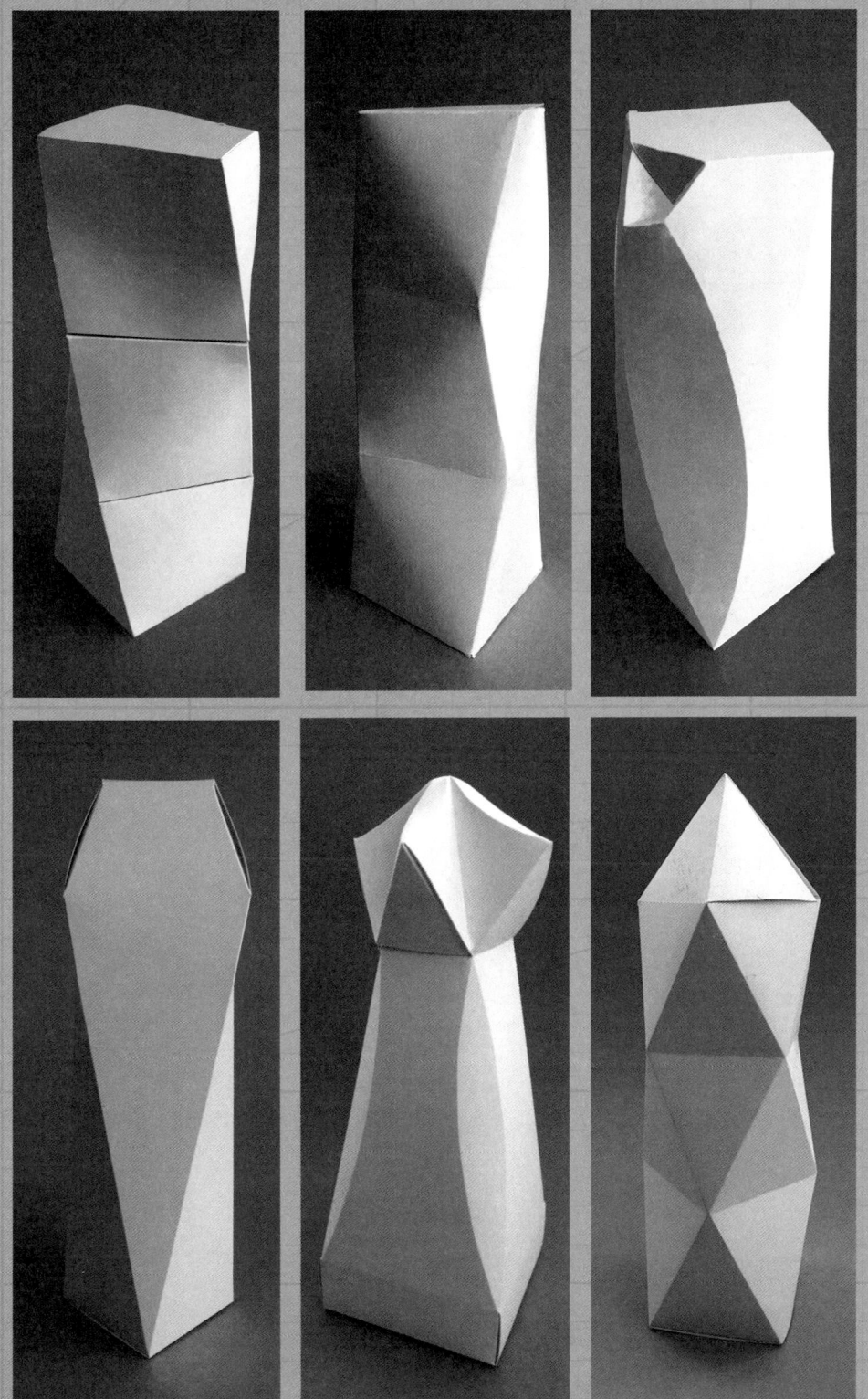

5.3 Beyond Packaging

The core of the net design system used throughout the book is to create a series of tabs on alternate edges around the perimeter of a net, and when folding the net from 2-D to 3-D, to slide the tabs under those edges which have no tabs. Done correctly, everything will lock securely.

This "tab, no tab, tab, no tab" interlocking system has many applications outside one-piece package design. These pages demonstrate a few of those possibilities.

5.3.1 Jazz Cube

This is similar to the simple Jackson Cube (see page 024), but much more decorative. It was designed by the author.

5.3.1.1

As a reminder, here is the square face of a cube. The "tab, no tab, tab, no tab" system can be applied to the 4 sides.

5.3.1.2

Tabs can be added to alternate edges to create the Jackson Cube. The Jazz Cube uses this principle.

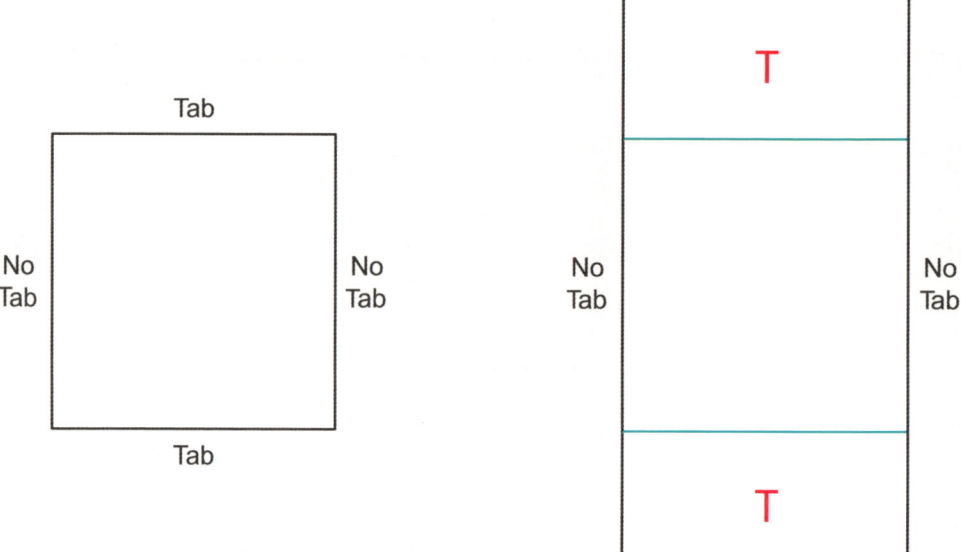

5.3.1.3

For preference, use a square of origami paper. Crease and unfold both diagonals. Fold a pair of opposite sides to the same diagonal.

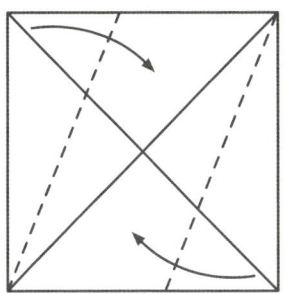

5.3.1.4

Turn over.

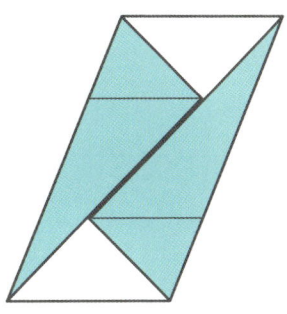

5.3.1.5

Fold dot to dot, twice, as shown.

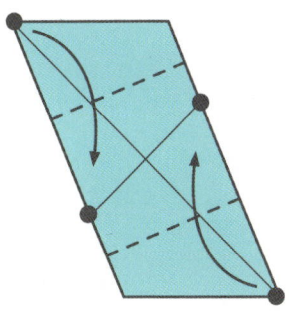

5.3.1.6

The result is a square. Unfold the last step.

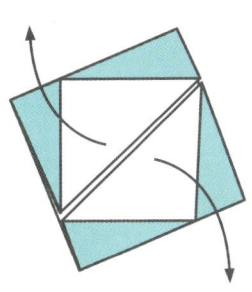

5.3.1.7

Turn over.

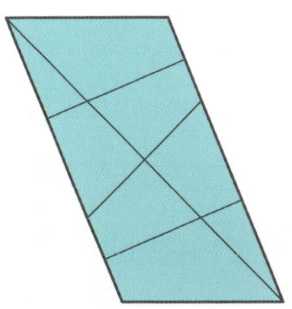

5.3.1.8

This is the completed unit.

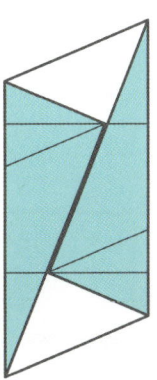

5.3.1.9

Note the central square and the trapezoid tabs, above and below.

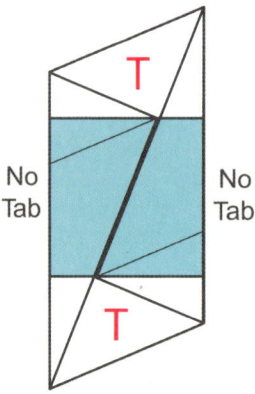

5.3.1.10

Make 6 units, being careful not to create any mirror symmetry units.

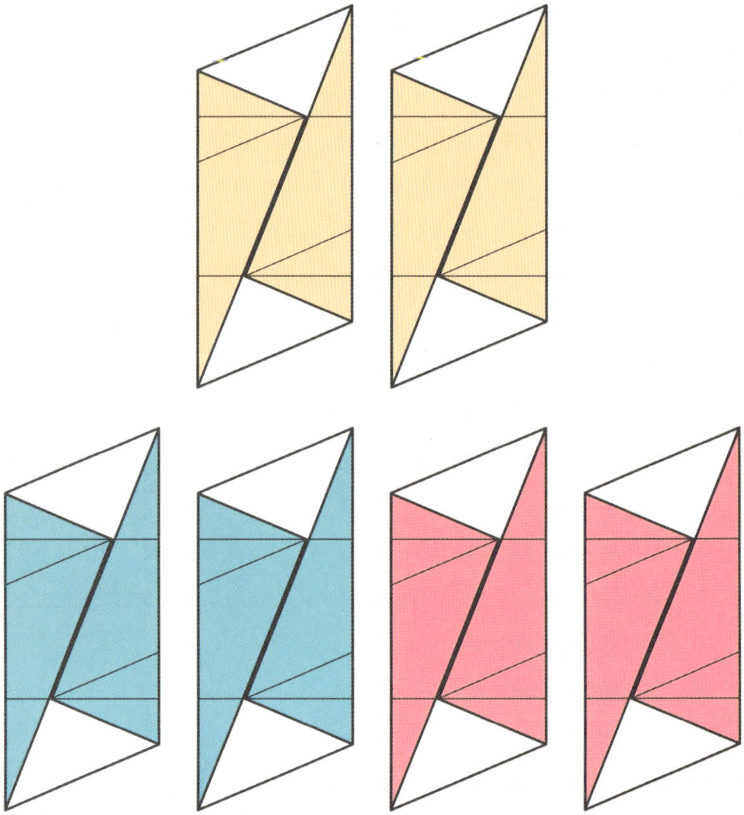

5.3.1.11

To lock the units together, hold 3 units as shown. Slide the acute corners of the red units under the blue triangles.

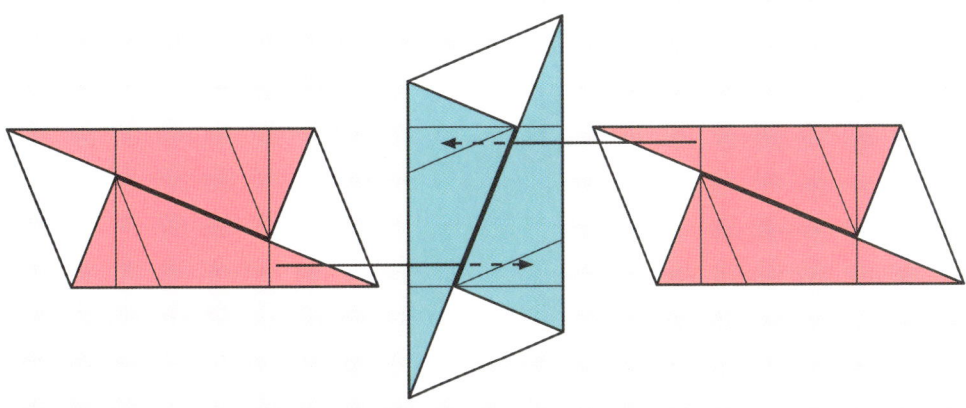

5.3.1.12

This is a completed face. The folds made in 5.3.1.5 must be folded at 90-degrees, to bend the units into the shape of a cube. Continue to lock the pieces together in the same way. At first, the cube will be weak, but when the final unit is interlocked, it will become strong.

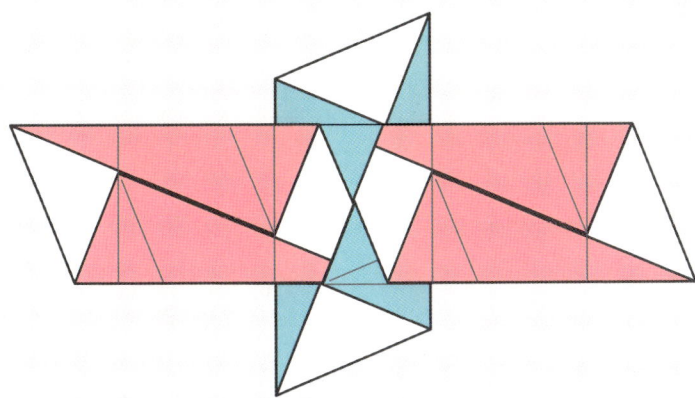

Note how the 12 tabs lock on the outside of the Cube, whereas the tabs on the Jackson Cube locked on the inside.

5.3.2 Harlequin Ball

This is another Jackson Cube variation. In this example, like the Jazz Cube above, the tabs again lock on the outside, but they are now triangular, not trapezoidal. This creates a different color pattern. Also, an additional fold in 5.3.2.11 allows the square face to divide into 2 triangles and thus, 12-units can be used to create a pyramided form, more complex than a 6-piece cube.

5.3.2.1

For preference, use a sheet of origami paper, white side up. Fold and unfold both diagonals.

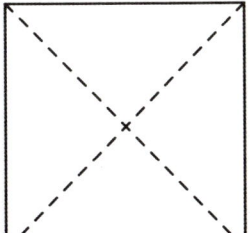

5.3.2.2

Turn the paper over.

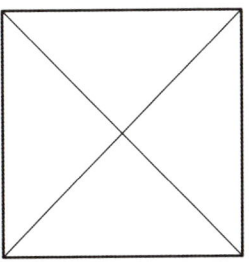

5.3.2.3

Fold a pair of opposite corners to the center point, then unfold.

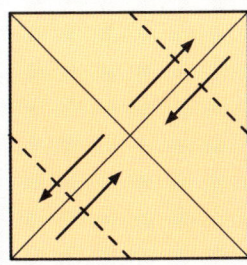

5.3.2.4

Turn over.

5.3.2.5

Fold the paper in half, as shown.

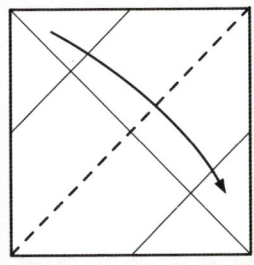

5.3.2.6

Carefully make a cut from the fold to the 5.3.2.3 folds.

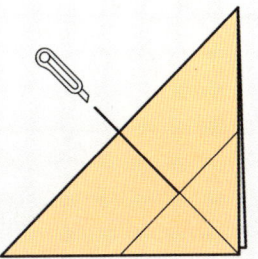

5.3.2.7

Spread A and B apart...

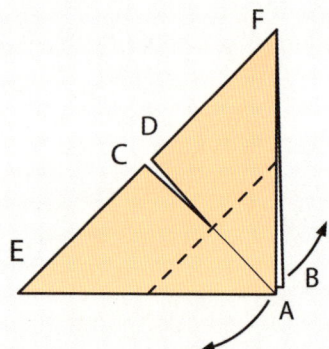

5.3.2.8

Like this. Flatten corner C to the left and D to the right... so that D touches B, and C touches A.

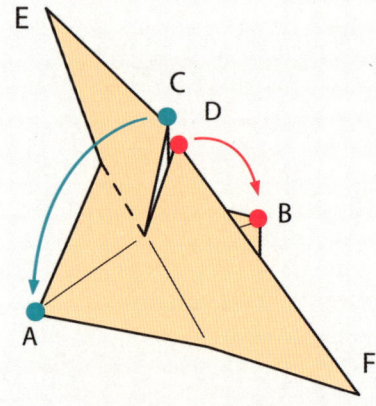

5.3.2.9

...halfway...

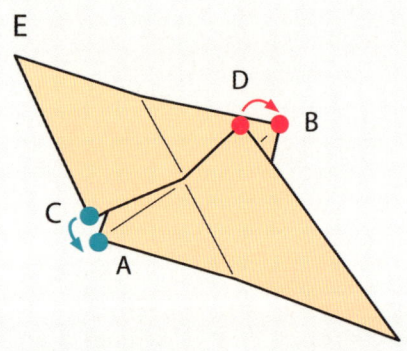

5.3.2.10

Turn over. The paper is flat.

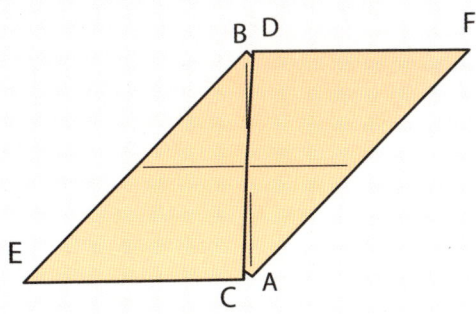

5.3.2.11

Make 1 valley fold and 2 mountain folds, as shown, so the unit becomes 3-D...

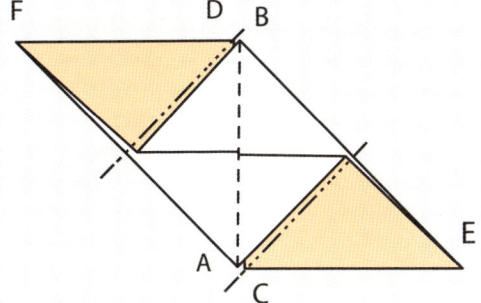

5.3.2.12

...like this. The colored triangles are the tabs.

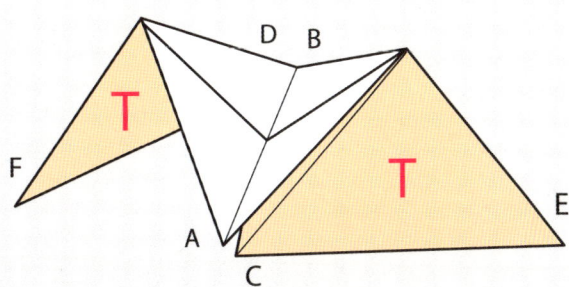

5.3.2.13

Make 12 units, being careful to not make any mirror images. Ideally, use 4 colors, making 3 units from each color.

5.3.2.14

This is the locking pattern that makes a 3-unit pyramid. The acute corner of a tab slides inside the white pocket of its neighbour. This happens 3-times to create a pyramid. Look carefully at the photograph. 4 pyramids meet at every corner. All 4 colors are present at each corner. With some planning, it is possible to make each of the harlequin squares consist of 2 white triangles and 2 triangles of only ONE color, not 2 different colors. Can you see how to do this?

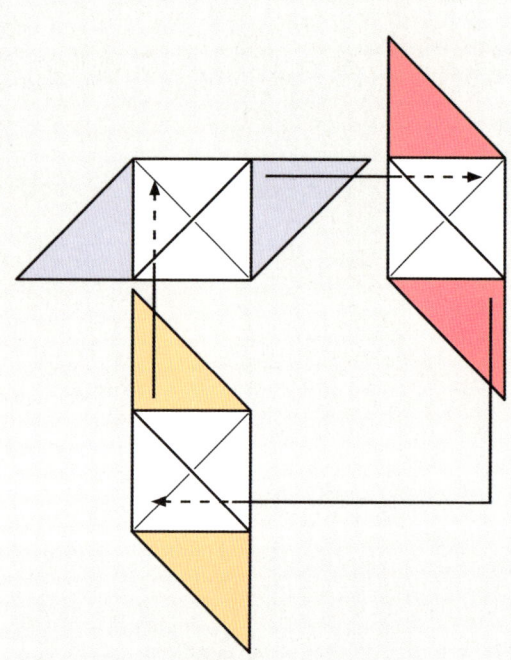

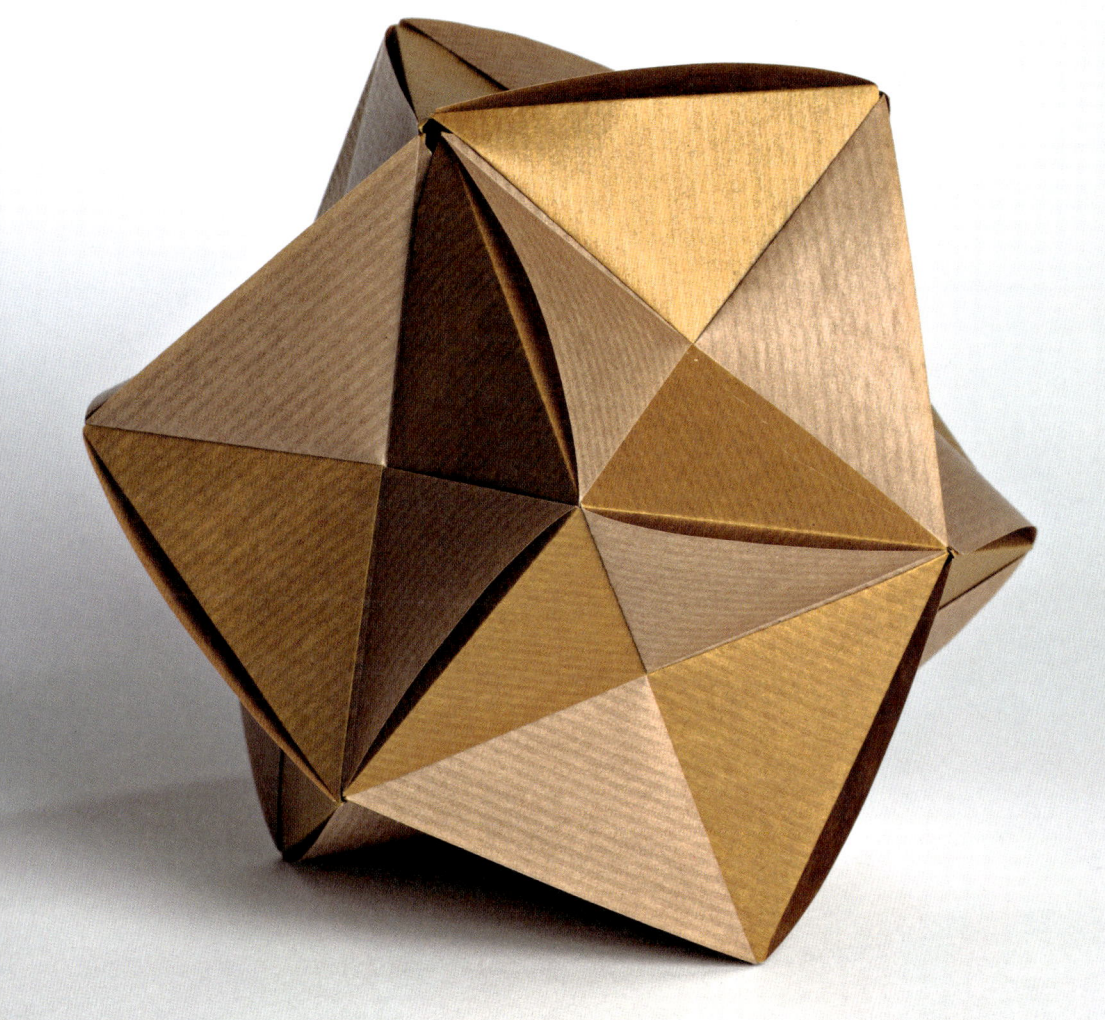

5.3.3 Half Tabbing

This is a system to make flat sheets of card interlock with no folding. Each tab occupies only half an edge.

5.3.3.1

A square has 4 sides, so the "tab, no tab, tab, no tab" system will work perfectly. It will also work for any polygon with an even number of sides, such as a hexagon or octagon.

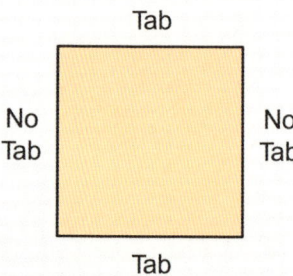

5.3.3.2

However, consider a polygon with an odd number of sides, such as a triangle. If one side is tabbed, the next will not have a tab…but what will happen to the third side? Will it be tabbed or untabbed? The system breaks down when we have an odd number of sides.

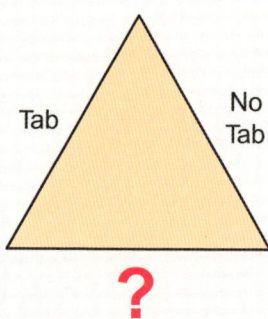

5.3.3.3

The solution is to half-tab each side, like this, dividing each side into a tabbed half and an untabbed half.

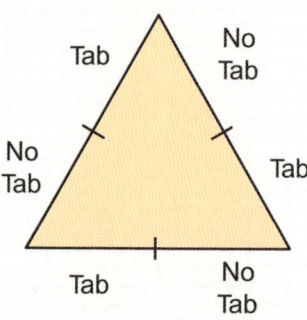

5.3.3.4

When the tabs are added, it can look like this.

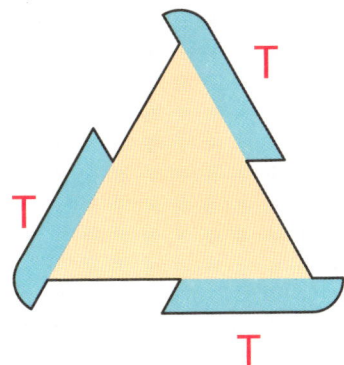

5.3.3.5

This is the half-tabbed unit. It does not need folding. The protruding quarter circles at the corners will hook around each other and the cut-back of the tab in the middle of each edge will help the tabs to hook tightly around each other.

Make the tabs from 250gsm card, not from paper.

5.3.3.6

2 units will lock as shown. Many 3-D polyhedrons can be made this way, using triangular units. If squares, pentagons, hexagons and other polygons, are made in a similar half-tabbed way and, crucially, they all have the same edge length, then they can be locked together to make a constellation of Platonic, Archimedean and other polyhedrons.

You can even build models of buildings, vehicles and other objects this way.

ACKNOWLEDGEMENTS

In the 35-year journey my net design system has so far undertaken, there are a huge number of people to thank. I must especially thank the many students who have unwittingly allowed me to test teaching ideas on them. They endured a sometimes eccentric project so I could indulge my fixation to teach the system as concisely and as memorably as possible. My motivation was to see the look of excitement and pride when a student had locked her/his first hand-made box!

The nets in Chapter 5 were designed by students at Aalto University, Helsinki, Finland, as part of the Pack Age module, supervised by Markus Joutsela and taught by me. The students are: Henrik Krogerus, Paula Honkonen, Sabina Friman, Amy Gelera, Annika Pöysti, Tuomas Mähönen and Atte Karppinen. The photographs of students' work were taken at Aalto University, Helsinki, Finland, Lahti University, Finland, the Hochschule für Gestaltung, Schwäbisch Gmünd, Germany and Brigham Young University-Idaho, Rexburg, Idaho, USA.

Creative Packaging:
One-Piece Packaging Solutions

Paul Jackson
©2022 Sendpoints Publishing Co., Ltd.
First printing of the first edition, May 2022

sendp●ints

PUBLISHED BY Sendpoints Publishing Co., Ltd.
ADDRESS: Unit 23, L1/F Mirror Tower, 61 Mody Road, Tsim Sha Tsui, Kowloon, Hong Kong, China
PUBLISHER: Lin Gengli
PUBLISHING DIRECTOR: Nicole Lo
CHIEF EDITOR: Nicole Lo
EXECUTIVE EDITOR: Huang Baomin
DESIGN DIRECTOR: Wu Dongyan
EXECUTIVE ART EDITOR: Ho Waikin
PROOFREADERS: Huang Chujun Chen Wenyin Zeng Wanting Liang Xinyi

SALES DIRECTOR: Philip Tsang
TEL: +852 6296 2246
EMAIL: sales@sppub.com
WEBSITE: www.sppub.com

ISBN 978-988-76087-5-2

All rights reserved. No part of this publication may be reproduced, stored in a retrieval system or transmitted in any form or by any means, electronic, mechanical, photocopying, recording or otherwise, without prior permission in writing from the publisher. For more information, please contact Sendpoints Publishing Co., Ltd.
Printed and bound in China.